SciencePlus®
TECHNOLOGY AND SOCIETY

LEVEL GREEN

Teaching Resources

Unit 2
Patterns of Living Things

HOLT, RINEHART AND WINSTON
Harcourt Brace & Company

Austin • New York • Orlando • Atlanta • San Francisco • Boston • Dallas • Toronto • London

To the Teacher

This booklet contains a comprehensive collection of teaching resources. You will find all of the blackline masters that you need to plan, implement, and assess this unit. Also included are worksheets that correspond directly to the SourceBook.

Choose from the following blackline masters to meet your needs and the needs of your students:

- **Home Connection** consists of a parent letter designed to get parents involved in the excitement of the *SciencePlus* method. The letter provides parents with a general idea of what you are going to cover in the unit, and it even gives you an opportunity to ask for any common household materials that you may need to accomplish the unit's activities most economically.
- **Discrepant Event Worksheets** provide demonstrations and activities that seem to challenge logic and reason. These worksheets motivate students to question their previous knowledge and to develop reasonable explanations for the discrepant phenomena.
- **Math Practice Worksheets** and **Graphing Practice Worksheets** help fine-tune math and graphing skills.
- **Theme Worksheets** encourage students to make connections among the major science disciplines.
- **Spanish Resources** include Spanish versions of the Home Connection letter, plus worksheets that outline the big ideas and principal vocabulary terms for the unit.
- **Transparency Worksheets** correspond to teaching transparencies to help you reteach, extend, or review major concepts.
- **SourceBook Activity Worksheets** reinforce content introduced in the SourceBook.
- **Resource Worksheets** consist of blackline-master versions of charts, graphs, and activities in the Pupil's Edition.
- **Exploration Worksheets** consist of blackline-master versions of Explorations in the Pupil's Edition. To help students focus on specific tasks, many of these worksheets include a goal, step-by-step instructions, and even cooperative-learning strategies. These worksheets simplify the tasks of assigning homework, allowing opportunities for make-up work, and providing lesson plans for substitute teachers.
- **Unit Activity Worksheet** consists of an activity, such as a crossword puzzle or word search, that provides a fun way for students to review vocabulary and main concepts.
- **Review Worksheets** consist of blackline-master versions of the review materials in the Pupil's Edition, including Challenge Your Thinking, Making Connections, and SourceBook Unit CheckUp.
- **Chapter Assessments** and **End-of-Unit Assessments** provide additional assessment questions. Each assessment worksheet includes two or more Challenge questions that encourage students to synthesize the main concepts of the chapter or unit and to apply them in their own lives.
- **Activity Assessments** are activity-based assessment worksheets that allow you to evaluate students' ability to solve problems using the tools, equipment, and techniques of science.
- **Self-Evaluation of Achievement** gives you an easy method of monitoring student progress by allowing students to evaluate themselves.
- **SourceBook Assessment** is an easy-to-grade test consisting of multiple-choice, true-false, and short-answer questions.

For your convenience, an **Answer Key** is available in the back of this booklet. The key includes reduced versions of all applicable worksheets, with answers included on each page.

Copyright © by Holt, Rinehart and Winston, Inc.

All rights reserved. No part of this publication may be reproduced or transmitted in any form or by any means, electronic or mechanical, including photocopy, recording, or any other information storage and retrieval system, without permission in writing from the publisher.

Permission is hereby granted to reproduce Student Worksheets and Activities in this publication in complete pages for instructional use and not for resale by any teacher using SCIENCEPLUS TECHNOLOGY AND SOCIETY.

Credits: See page 90.

SCIENCEPLUS is a registered trademark of Harcourt Brace & Company licensed to Holt, Rinehart and Winston, Inc.

Printed in the United States of America

ISBN 0-03-095660-9

4 5 6 7 8 9 021 06 05 04 03 02 01

Unit 2: Patterns of Living Things

Contents

Worksheet	Page	Suggested Point of Use
Home Connection	1	Prior to Unit 2
Chapter 4		
Exploration Worksheet ▼	3	Lesson 2, Exploration 1: Observing Locomotion, p. 66
Transparency Worksheet ▼	6	Lesson 2, Animal Innovators, p. 67
Exploration Worksheet	8	Lesson 3, Exploration 2: Solve the Insect Movement Mystery, p. 69
Exploration Worksheet ▼	9	Lesson 3, Exploration 3: How's Your Horse Sense? p. 69
Exploration Worksheet	11	Lesson 3, Exploration 4: Listening to an Earthworm! p. 72
Review Worksheet	13	Challenge Your Thinking, p. 73
Chapter 4 Assessment	16	End of Chapter 4
Chapter 5		
Activity Worksheet: Growth Trivia	19	Lesson 2, p. 80
Exploration Worksheet	20	Lesson 2, Exploration 1: Changing Conditions, p. 81
Graphing Practice Worksheet	21	End of Lesson 2
Activity Worksheet: Growth Puzzle ▼	23	End of Lesson 2
Review Worksheet ▼	25	Challenge Your Thinking, p. 89
Chapter 5 Assessment	29	End of Chapter 5
Chapter 6		
Exploration Worksheet	32	Lesson 1, Exploration 1: Experimenting With Stimulus and Response: An Earthworm Responds, p. 92
Discrepant Event Worksheet	34	Prior to Lesson 3
Exploration Worksheet ▼	35	Lesson 3, Exploration 3: Which Loses Heat Faster— A Mouse or a Mountain Lion? p. 103
Activity Worksheet: A Warmblooded Puzzle	36	End of Lesson 3
Theme Worksheet	37	Lesson 4, Plants and Animals Made to Order, p. 109
Review Worksheet	39	Challenge Your Thinking, p. 110
Chapter 6 Assessment	42	End of Chapter 6
Chapter 7		
Resource Worksheet ▼	44	Lesson 1, How to Use a Microscope, p. 114
Transparency Worksheet ▼	45	Lesson 1, How to Use a Microscope, p. 114
Exploration Worksheet	47	Lesson 1, Exploration 1: Looking at Cells, p. 115
Transparency Worksheet ▼	48	Lesson 2, Skilled Observations of Cells, p. 116
Exploration Worksheet	50	Lesson 2, Exploration 2: What Are Plants and Animals Made Of? p. 116
Exploration Worksheet	52	Lesson 2, Exploration 3: Edible Cells, p. 118
Review Worksheet	54	Challenge Your Thinking, p. 120
Chapter 7 Assessment	57	End of Chapter 7

▼ *A corresponding transparency is available. See the Teaching Transparencies Cross-Reference on page v.*

Contents, continued

Worksheet	Page	Suggested Point of Use
Unit Review and Assessment		
Unit Review:		
Unit Activity Worksheet: A Magic Square	59	End of Unit 2
Unit Review Worksheet	60	Making Connections, p. 122
Unit Assessment:		
End-of-Unit Assessment	64	End of Unit 2
Activity Assessment	69	End of Unit 2
Self-Evaluation of Achievement	72	End of Unit 2
SourceBook		
SourceBook Activity Worksheet	73	SourceBook, p. S24
SourceBook Review Worksheet	75	SourceBook, Unit CheckUp, p. S63
SourceBook Assessment	79	End of SourceBook Unit 2
Spanish		
Contacto en la casa (Home Connection)	85	Prior to Unit 2
Las grandes ideas (The Big Ideas)	87	Prior to Unit 2/Throughout Unit 2
Vocabulario (Vocabulary)	88	Prior to Unit 2/Throughout Unit 2
Answer Keys		
Chapter 4	90	Throughout Chapter 4
Chapter 5	96	Throughout Chapter 5
Chapter 6	102	Throughout Chapter 6
Chapter 7	106	Throughout Chapter 7
Unit 2	111	End of Unit 2
SourceBook Unit 2	117	End of SourceBook Unit 2

Teaching Transparencies Cross-Reference

Transparency	Corresponding Worksheet	Suggested Point of Use
Chapter 4		
8: Observing Locomotion Table	Exploration Worksheet, p. 3	Lesson 2, Exploration 1: Observing Locomotion, p. 66
9: Frog Metamorphosis	Transparency Worksheet, p. 6	Lesson 2, Animal Innovators, p. 67
10: How's Your Horse Sense?	Exploration Worksheet, p. 9	Lesson 3, Exploration 3: How's Your Horse Sense? p. 69
Chapter 5		
11: Human Growth Stages		Lesson 1, p. 78
12: Growth Puzzle	Activity Worksheet, p. 23	End of Lesson 2
13: Class Growth Graph	Review Worksheet, p. 26	Challenge Your Thinking, p. 89
Chapter 6		
14: Body Temperatures Graph	Exploration Worksheet, p. 35	Lesson 3, Exploration 3: Which Loses Heat Faster—A Mouse or a Mountain Lion? p. 103
15: Bird Adaptations		Lesson 4, p. 106
Chapter 7		
16: Using a Microscope	Resource Worksheet, p. 44	Lesson 1, How to Use a Microscope, p. 114
17: Compound Light Microscope	Transparency Worksheet, p. 45	Lesson 1, How to Use a Microscope, p. 114
18: Look and See		Lesson 1, p. 113
19: A Good Drawing		Lesson 1, p. 113
20: Plant and Animal Cells	Transparency Worksheet, p. 48	Lesson 2, Skilled Observations of Cells, p. 116
21: Different Kinds of Cells		Lesson 2, Plant or Animal? p. 119

SCIENCEPLUS • LEVEL GREEN

**Unit 2
Home Connection**

Dear Parent,

In the next few weeks, your son or daughter will be actively investigating some of the characteristics of living things, namely, the ability to move, to grow, and to respond to stimuli. By the time students have finished Unit 2, they should be able to answer the following questions to grasp the "big ideas" of the unit.

1. What are some signs of life? (Ch. 4)
2. How do plant and animal movement differ? (Ch. 4)
3. In what ways are the growth patterns of plants and people similar? (Ch. 5)
4. In what ways are they different? (Ch. 5)
5. What are some of the different forms of growth? (Ch. 5)
6. What is the importance of each? (Ch. 5)
7. What does "response to stimuli" mean? (Ch. 6)
8. Why is this response valuable? (Ch. 6)
9. What evidence is there that animals have a sense of time? (Ch. 6)
10. What are some adaptations of plants and animals over long periods of time? (Ch. 6)
11. How do plant and animal species adapt over long periods of time? (Ch. 6)
12. How are plant and animal cells similar? How do they differ? (Ch. 7)
13. Describe the procedure for making a wet mount. (Ch. 7)

Listed below are some activities that you may want to do with your son or daughter at home.

- Have your son or daughter find things around the house that are living and things that are not living. Ask your child what similarities and differences he or she notices about these two categories of things. Can he or she find things that are not living now but were once living?
- Discuss with your child his or her growth patterns. Try to recount the growth spurts he or she may have had. Use family photographs as visual aids to illustrate the changes. Compare your son's or daughter's growth patterns with what you know about your own growth patterns as you grew up.

Sincerely,

Unit 2
Home Connection

The items listed below are materials that we will use in class for the various science explorations of Unit 2. Your contribution of materials would be very much appreciated. I have checked certain items below. If you have these items and are willing to donate them, please send them to the school with your son or daughter by

_____.

- ○ cotton
- ○ cups (small, plastic)
- ○ dark-colored cloth or cardboard
- ○ eyedroppers
- ○ flashlights
- ○ forceps
- ○ glass bottles with lids
- ○ glass jars (with lids; small)
- ○ graph paper

- ○ measuring tapes (metric)
- ○ newspaper
- ○ oven mitts
- ○ paper towels
- ○ pitchers
- ○ plastic wrap
- ○ rubber bands (large)
- ○ sand
- ○ sawdust

- ○ seeds (quick-growing, such as mung bean, alfalfa, radish, carrot, or mustard)
- ○ shallow pans or baking sheets
- ○ shoe boxes
- ○ soil
- ○ straight pins
- ○ tissue

Thank you in advance for your help.

Name _____ Date _____ Class _____

EXPLORATION 1

Chapter 4
Exploration Worksheet

Observing Locomotion, page 66

Cooperative Learning Activity		**Safety Alert!**
Group size	3 to 4 students	Be very gentle with the animals so that you do not harm them. If possible, return each animal to its natural environment at the end of the Exploration.
Group goal	to analyze the locomotion of several animals	
Individual responsibility	Each member of your group should choose a role such as primary investigator, recorder, pacer, or materials coordinator.	
Individual accountability	Each group member should be able to summarize your group's findings and to rate his or her own performance.	

You Will Need

- a shoe box
- a large rubber band
- a cloth or piece of cardboard
- several small animals
- plastic wrap
- a pin
- a watch or clock

What to Do

For this experiment, you can use any of the following small animals: a spider, an ant, a sow bug, a grasshopper, an earthworm, a caterpillar, a millipede, or a lizard. Be very gentle so that you do not harm the animals. Record your observations in the table on the next page.

1. Put one of the animals in a shoe box or other small container, and cover the container with plastic wrap.
2. Poke several small holes in the wrap with a pin.
3. Cover the container with something that will keep out light. (A cloth or piece of cardboard will do.)
4. Remove the covering carefully, and watch the animal closely. How does the animal move when the cover is first removed? Record your observations in the table on the next page.
5. Continue to observe the animal for 2 minutes. Clearly and in as much detail as possible, describe the way the animal moves. Estimate how far the animal moves during the 2 minutes. Record your observations in the table on the next page.
6. Repeat the experiment using other animals, and compare the results.
7. Using your data, fill out a lab report that includes the information in the table.
8. Wash your hands after working with any animal. When you have finished the Exploration, be sure to return the animal to the place you found it.

SCIENCEPLUS • LEVEL GREEN 3

Name _____ Date _____ Class _____

Exploration 1 Worksheet, continued

Name of animal	How the animal moved when the cover was first removed	Description of animal movement	How far the animal moved in 2 minutes

Name _____ Date _____ Class _____

Exploration 1 Worksheet, continued

Questions

1. What do you think made the animals move when the cover was first removed?

   _____
   _____
   _____

2. Why do you think the animals continued to move?

   _____
   _____
   _____

3. Which body parts did each animal use for locomotion?

   _____
   _____
   _____

4. If all of the animals you observed were in a 2-minute race, which one would win?

   _____
   _____
   _____

5. What characteristic of this animal would make it the winner of such a race?

   _____
   _____

Photos also on page 68 of your textbook

6. According to your data, which animal would come in last? Can you suggest reasons why?

   _____
   _____
   _____
   _____
   _____

SCIENCEPLUS • LEVEL GREEN

Name _____ Date _____ Class _____

Chapter 4
Transparency Worksheet

Frog Metamorphosis Teacher's Notes

This worksheet corresponds to Transparency 9 in the Teaching Transparencies binder.

Suggested Uses

Use the transparency as a visual aid with any of the topics listed below.
  Animal Innovators, page 67
  Reproduction, page 87
  Adaptation of Structure, page 105

Use the transparency to extend the study of locomotion, reproduction, and adaptation.

Use the transparency with the transparency worksheet on the next page for reteaching or review. Please note: The worksheet is a reproduction of the actual transparency with certain labels omitted. For answers to that worksheet, see the transparency.

Possible Extension Questions

1. What are some human inventions that seem to imitate the frog's method of movement?

2. How does the breathing of the adult frog differ from the breathing of the tadpole?

3. Explain how the frog has adapted to its environment.

Answers to Extension Questions

1. Answers will vary. Possible answers may include comparisons between the frog's webbed feet and human diving fins or flippers, and between the tail of the tadpole and paddles or oars.

2. The adult frog uses lungs for breathing, while the tadpole uses gills.

3. Answers will vary but should focus on the fact that the frog is adapted to live both on land and in water. In connection with this, the students may describe features such as the frog's long, sticky tongue, webbed feet, leg structure, body coloration, eyelids, the way its eyes are set atop its head, etc. Also, students should be able to explain adaptations specific to the metamorphic tadpole stages.

Name _____ Date _____ Class _____

Frog Metamorphosis

Chapter 4
Transparency Worksheet

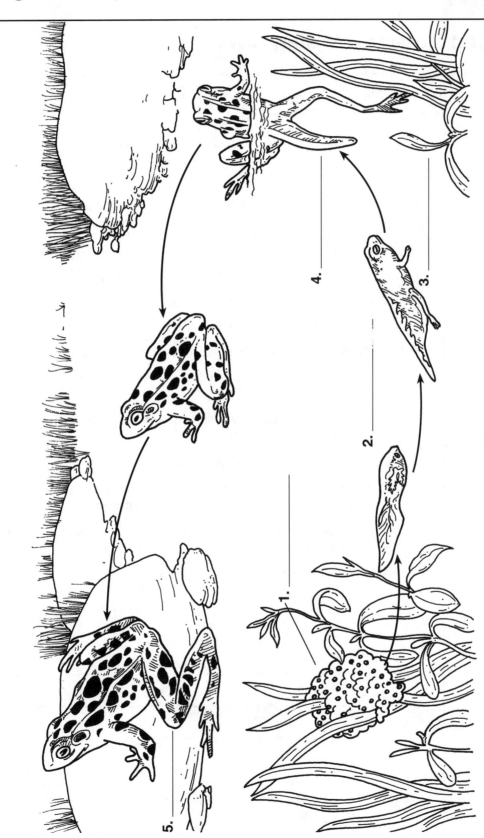

Name _____ Date _____ Class _____

EXPLORATION 2

Chapter 4
Exploration Worksheet

Solve the Insect Movement Mystery, page 69

Your goal	to learn more about the process of locomotion by determining how an insect moves	Safety Alert!

You Will Need

- a crawling insect
- a piece of paper

What to Do

1. Put a crawling insect on a piece of paper.
2. Watch carefully to see how the insect moves. Decide which of the following things happens:

    a. Both of the front legs move forward at the same time.

    b. Both of the middle legs move forward at the same time.

    c. Both of the back legs move forward at the same time.

    d. All three legs on one side move forward at the same time.

    e. On one side, the front and middle legs move forward at the same time.

    f. On one side, the front and back legs move forward at the same time.

    g. All of the legs move at different times.

    h. Three legs move forward at the same time: the front and middle legs on one side and the back leg on the other side.

    i. Three legs move forward at the same time: the front and back legs on one side and the middle leg on the other side.

3. Record your choice and why you think it is correct.

   _____
   _____
   _____
   _____
   _____
   _____

8 UNIT 2 • PATTERNS OF LIVING THINGS

Name _____ Date _____ Class _____

EXPLORATION 3

**Chapter 4
Exploration Worksheet**

How's Your Horse Sense? page 69

| **Your goal** | to learn more about the process of locomotion by determining how a horse moves |

Below (and on page 69 of your textbook) is a sequence of pictures showing a horse walking. The pictures are arranged to show the order in which steps are taken. Study the set of pictures carefully. Then answer the following questions.

Illustration also on page 69 of your textbook

Questions

1. As the horse walks, how many hooves touch the ground at a time? Are all of the horse's hooves off the ground at any one time?

   _____
   _____
   _____

2. Describe the way the horse moves its legs throughout the walking sequence—for example, "right foreleg first, left hind leg second," and so on.

   _____
   _____
   _____

3. Have you ever seen a baby crawl? How does the horse's walking gait compare to the movements of a crawling baby?

   _____
   _____
   _____

SCIENCEPLUS • LEVEL GREEN 9

Name _____ Date _____ Class _____

Exploration 3 Worksheet, continued

Comparing Structures

1. The illustrations below show a horse's forelimb and its human counterpart. Label the equivalents of the fingernails, fingers, hand, wrist, forearm, elbow, and upper arm on the diagram of the horse's leg.

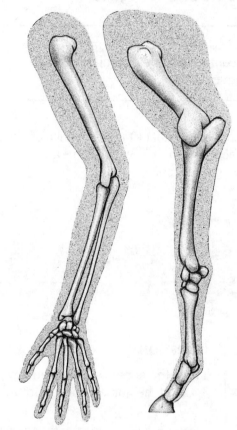

Illustration also on page 70 of your textbook

2. How is each structure in each forelimb well suited to its function?

_____
_____
_____
_____
_____
_____
_____
_____

UNIT 2 • PATTERNS OF LIVING THINGS

Name _____ Date _____ Class _____

EXPLORATION 4

Chapter 4
Exploration Worksheet

Listening to an Earthworm! page 72

| **Your goal** | to determine how an earthworm moves |

You Will Need

- an earthworm
- a piece of stiff paper
- 10 mL of water

What to Do

1. Put an earthworm on a piece of stiff paper. Can you hear it make any noise as it moves?

2. Hold the paper up level with your eyes, and try to look between the animal and the paper. Do you see the bristles? Rub your finger back and forth along the lower side of the earthworm. Do you feel the bristles? When the stiff bristles are extended, they hold the animal in place on the ground. When the bristles are pulled in, the animal can slide along. When the earthworm moves along the paper, its bristles sometimes make a scraping noise.

3. Notice the rings on the body of the worm. The inside of an earthworm's body is divided into sections called *segments*, which show up on the outside as rings. There are four pairs of bristles for every segment. How many bristles does your earthworm have?

   _____

4. Now put the earthworm on damp paper. Watch it move. How does the earthworm use its bristles to propel itself?

   _____

   _____

   Like snakes, earthworms have muscles. How does the earthworm use its muscles to move?

   _____

   _____

An inside and lengthwise look at an earthworm

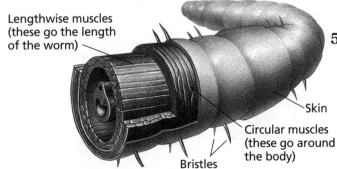

Lengthwise muscles (these go the length of the worm)
Skin
Circular muscles (these go around the body)
Bristles

5. Look at the illustration. Locate the muscles. These muscles are very strong. One set of muscles goes around the worm in rings.

Illustration also on page 72 of your textbook

SCIENCEPLUS • LEVEL GREEN **11**

Name _____ Date _____ Class _____

Exploration 4 Worksheet, continued

6. Now locate the muscles that run lengthwise along the worm's body. The two sets of muscles work against each other.

   a. What happens when one set of these muscles contracts and the other set relaxes?

   _____
   _____
   _____

   b. When the ring muscles contract, what happens to the length of the segment?

   _____
   _____
   _____

   c. When the lengthwise muscles contract, what happens to the length of the segment?

   _____
   _____
   _____

7. See if you can piece together this information about the muscles and bristles to write a description of how earthworms move.

   _____
   _____
   _____
   _____

8. Compare your description with Dan's description on page 72 of your textbook.

   _____
   _____
   _____
   _____
   _____

Name _____ Date _____ Class _____

Chapter 4
Review Worksheet

Challenge Your Thinking, page 73

1. Wanted: Nonliving or Living

What characteristics do all living things have in common?

_____
_____
_____

Sometimes a nonliving thing has a characteristic of a living thing. How many examples of this can you list? Make a list. Can you top 20 examples?

_____
_____
_____
_____
_____
_____

If a nonliving thing has a characteristic of something that is alive, why is it classified as nonliving?

_____
_____
_____
_____

2. Classify It

A biologist made the following classification of all objects:

a. living—having all of the signs of life

b. dead—having once had all of the signs of life

c. nonliving—having never had all of the signs of life

Where does a wooden chair fit in? How about other things—water, mushrooms, a pie? Classify each of the following into one of the three groups: oyster, moss, yeast cake, salt, sugar, bones, hibernating bear, pencil, wool sweater, seaweed, sponge, pearl, volcano, kernel of corn, clams, barnacle on a rock, bean seed, baked beans, freshly picked strawberries, pine cone, lichens on a rock, electric fan, cactus, paper.

Record your answers in the chart on the next page.

SCIENCEPLUS • LEVEL GREEN 13

Chapter 4 Review Worksheet, continued

Living	Dead	Nonliving

3. Curious Questions

Your class has just been asked to write the section called "Curious Questions" in the book *Mr. Know-It-All's Science Facts*. Choose one question from the list below. The question should be answered clearly in a paragraph of 150 words or less and should be understandable to an 11- or 12-year-old. If an illustration would help, draw one! Continue in your ScienceLog if necessary.

a. How do you tell a plant from an animal?

b. In how many different ways can animals move?

c. How does an insect use its six legs to move?

d. How does a horse use its four legs to move?

_____
_____
_____
_____
_____
_____
_____
_____
_____
_____
_____

Name _____ Date _____ Class _____

Chapter 4 Review Worksheet, continued

4. That's Life? You are a scientific explorer, and you have just come across the mysterious blob shown on page 74 of your textbook. Do you think it is alive? What characteristics would you look for? Write down a series of steps you would take to answer these questions.

_____

_____

_____

_____

_____

_____

_____

5. It's All Connected Create a concept map to show how the following words and phrases are related: *living things, movement, human, scales, bristles, limbs, snake, animal,* and *earthworm.* To create the map, arrange these items in a logical way, and use lines and connecting phrases to link them.

Name _____ Date _____ Class _____

Chapter 4
Assessment

Short Responses

1. Do you agree or disagree with the following statement? Explain your reasoning.

   Plants cannot move because they do not have the necessary limbs for locomotion.

   _____
   _____
   _____
   _____

2. Explain why food production is not one of the universal signs of life.

   _____
   _____
   _____
   _____

Illustration for Interpretation

3. Use the illustration below to answer the questions on the next page.

16 UNIT 2 • PATTERNS OF LIVING THINGS

Name _____ Date _____ Class _____

Chapter 4 Assessment, continued

a. List the living, nonliving, and dead things that you see in the illustration.

Living	
Nonliving	
Dead	

b. What characteristics of living things help you distinguish them from nonliving things?

_____
_____

CHALLENGE 1
Short Essay

4. Earthworms and snakes have similar means of locomotion. Compare and contrast each animal's method of getting from place to place, describing the techniques of each one and pointing out their similarities and differences.

_____
_____
_____
_____
_____
_____
_____
_____
_____

Name _____ Date _____ Class _____

Chapter 4 Assessment, continued

Numerical Problem

5. An earthworm has 4 pairs of bristles on each of its segments. Suppose that you find an earthworm with 64 segments.

   a. How many individual bristles does the earthworm have? Show your work.

   b. You find another earthworm that is three-fourths of the size of the first earthworm. How many individual bristles does this earthworm have? Show your work.

   c. Why might a scientist be interested in the number of bristles on an earthworm?

18 UNIT 2 • PATTERNS OF LIVING THINGS

Name _____ Date _____ Class _____

Chapter 5
Activity Worksheet

Growth Trivia

This activity corresponds to Lesson 2, which begins on page 80 of your textbook.

Who can find the most answers? Give yourself 48 hours to answer the following questions:

1. How long does it take a puppy to become full grown?	
2. Does the life span of an animal have anything to do with how long it takes it to grow up?	
3. A dog at 14 years of age is physically as old as a human at 98 years. How many years of a human's life equals 1 year of a dog's life?	
4. How long does a baby elephant remain dependent on its mother?	
5. At what age is an elephant full grown?	
6. On average, how long does an elephant live?	
7. Name two mammals that grow hair only after they are born.	
8. List two animals that, when they are born, are basically on their own.	
9. How does a hen show signs of old age?	
10. How does a lobster or a grasshopper—both of which are covered by hard shells—grow?	
11. What animals change completely as they grow, so that the young do not look like the adults?	
12. How are chickens different from each other— even those that look alike?	
13. How many hairs are on your head?	
14. As we grow, we often grow differently; but what is different about every human, even at birth?	
15. How tall is, or was, the tallest human?	

SCIENCEPLUS • LEVEL GREEN   19

Name _____ Date _____ Class _____

EXPLORATION 1

Chapter 5
Exploration Worksheet

Changing Conditions, page 81

Cooperative Learning Activity	
Group size	3 to 4 students
Group goal	to test different conditions for growing seeds and to develop an understanding of the term *variable*
Individual responsibility	Each member of your group should choose a role such as materials coordinator, recorder, leader, or investigator.
Individual accountability	Each group member should be able to define the term *variable* and explain why it is important to identify variables when conducting an experiment.

Imagine that you are a botanist working for a seed company. You are responsible for testing different conditions for growing seeds and for developing directions to go on the label.

You Will Need

- 10–20 quick-growing seeds, such as mung bean, alfalfa, radish, carrot, or mustard seeds
- a variety of seed-growing materials provided by your teacher

What to Do

1. Divide your seeds into several small groups. You should have more than one seed in each group.

2. Discover what conditions are needed to produce the best results.

3. In your ScienceLog, make a report for the company files, and write a draft of the instructions for the seed package. In your report, include the following information: the sets of conditions you tested, what happened to each group of seeds, and which conditions were the best.

4. One word you might find useful in writing the report is *germinate*. This word means "to begin to grow from a seed."

5. In performing your tests, be sure to change only one variable. Remember that a variable is a condition that can or does change. Variables that you could change include temperature, moisture, and amount of light. All other variables should remain the same.

6. Some methods for germinating seeds are shown on page 81 of your textbook. Use the same type of seed for each set of conditions. Try to include bean, carrot, or radish seeds. You can read about the method one group of students used on page 81. They used several seeds for each condition. Why is this important?

   _____

7. Before you begin, make a prediction in your ScienceLog about what you think the result of each experiment will be.

20   UNIT 2 • PATTERNS OF LIVING THINGS

Name _____ Date _____ Class _____

**Chapter 5
Graphing Practice Worksheet**

Do You Have a Green Thumb?

Do this activity after completing Lesson 2, which begins on page 80 of your textbook.

Jane has two fields in different parts of the country, and she wants to plant seeds and raise crops on them. In preparation for this task, Jane did a little research. First, she used an almanac to look up the average temperature of each region throughout the year. Then she graphed her findings so they would be easier to read.

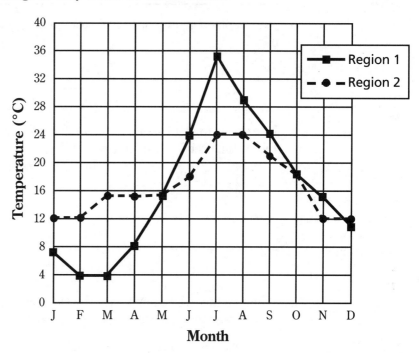

Next, Jane bought three types of plant seeds and labeled them *A*, *B*, and *C*. Each type of seed has a temperature at which it grows best. Each type of seed also has a minimum temperature for growth, below which the seed freezes, and a maximum temperature for growth, above which the seed shrivels. Jane knows she needs to be careful when and where she plants the seeds so that she can be sure to grow healthy plants. So she looked up the minimum, maximum, and ideal temperatures for each seed and listed them in the table below.

Seed type	Min. temp. (°C)	Ideal temp. (°C)	Max. temp. (°C)
A	7	15	26
B	1	7	18
C	12	28	40

SCIENCEPLUS • LEVEL GREEN 21

Name _____ Date _____ Class _____

Graphing Practice Worksheet, continued

Questions Jane is ready to begin planting. Can you help her? Use the graph to answer the following questions:

1. Will any of the seeds freeze in region 1? in region 2?

_____

_____

2. Will any of the seeds shrivel in region 1? in region 2?

_____

_____

3. Jane wants to grow plants all year long in region 1. Which seed should she use?

_____

_____

4. Jane wants to plant seeds in region 2 in March and harvest the plants in July. Which seed should she use? Why?

_____

_____

_____

_____

5. When do the two regions have the same average temperature? When do the average temperatures differ the most?

_____

_____

_____

6. List at least four factors other than temperature that might influence how well Jane's plants grow.

_____

_____

_____

_____

_____

Name _____ Date _____ Class _____

Growth Puzzle

Tackle this crossword puzzle after you've finished Lesson 2, which begins on page 80 of your textbook.

Chapter 5 Activity Worksheet

Each paragraph on the following page is a summary of material you have studied so far. You will notice, though, that key words are missing. In their place is a number followed by the word *across* or *down*. When you think you have the right word for a blank, see whether your answer fits into the boxes for that number across or down.

Growth Puzzle, continued

Summaries for Completion

A. There are *(2, down)* in growth or growth rates in any group of living things. In humans, the *(5, across)* is the part of the body that develops the earliest in life. In a human baby before birth, the length of the *(18, down)* is small in proportion to the length of the whole body. At birth, the *(1, down)* is the smallest in proportion to its eventual adult size. The *(10, across)* of an adult male is roughly twice what it was at 2 years of age.

B. In young plants, *(19, across)* develop faster than stems. Two important conditions for the growth of plants from seeds are *(13, across)* and *(12, down)*. When a seed begins to grow, we say that it sprouts, or *(11, down)*. In performing growth experiments with plants, changing certain conditions affects their growth. A condition that we change is called a *(4, down)*. If we plant several seeds 2 cm below the surface of the soil and plant other seeds of the same kind 4 cm below the surface, we are investigating the effect of *(6, down)* on seed growth.

C. Animals that are large in size experience a number of problems. One is their need for more *(16, across)*.

D. The age of living things can often be discovered from growth patterns, such as those in the *(23, across)* of a tree. The growth of our *(7, across)* and *(9, across)* is an example of continuous growth. The teeth of certain animals, such as *(22, across)*, grow continually but are kept short by gnawing.

E. When some animals lose a body part, they have the ability to *(15, across)* this part. For example, a *(3, across)* can lose an arm and grow a new one. The same is true of the claw of a *(19, down)*. The sea cucumber, when attacked, discharges its internal *(17, down)* and then replaces them later. When we cut ourselves, the *(8, across)* of the injured part is a kind of growth. When a plant or animal forms another organism like itself, this type of activity is called *(14, down)*.

F. Sometimes wild, uncontrolled growth called *(20, across)* occurs in our bodies. Another kind of growth or enlargement may occur in a plant; an example is the *(21, down)* on a goldenrod.

Name _____ Date _____ Class _____

**Chapter 5
Review Worksheet**

Challenge Your Thinking, page 89

1. An Eggsercise   Take a look at the photograph of the chicken egg below. Use the photograph and what you have learned in this chapter to answer the following questions:

Photo also on page 89 of your textbook

a. What is the purpose of the eggshell?

_____

_____

b. What is the purpose of the yolk? How long does it have to last?

_____

_____

c. Where in the egg would the young chick have developed if the egg had been fertilized?

_____

_____

d. How does the structure of an egg protect a developing chick?

_____

_____

_____

_____

_____

SCIENCEPLUS • LEVEL GREEN   25

Name _____ Date _____ Class _____

Chapter 5 Review Worksheet, continued

2. Check It Out — Look at the bar graph illustrating the height of students in Ada's class, and then answer the following questions. Ada is 130 cm tall.

  a. How many students are in Ada's class? _____

  b. How many boys are shorter than Ada? _____

  c. How many girls are as tall as Ada or taller? _____

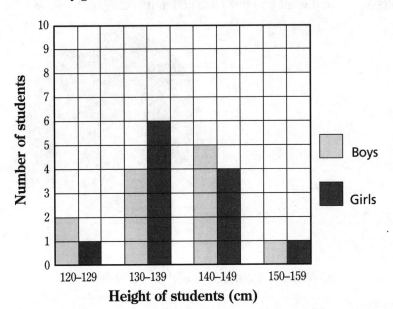

3. Inquiring Minds — Add the answers to the following "Curious Questions" that you worked on earlier in the unit.

  a. Why does a baby have such a big head?

  _____
  _____
  _____
  _____

  b. Is bigger always better?

  _____
  _____
  _____
  _____

Name _____ Date _____ Class _____

Chapter 5 Review Worksheet, continued

    c. How can you tell the ages of plants and animals?

_____

_____

_____

_____

_____

    d. Can an animal lose a part of its body and grow a new one to replace it?

_____

_____

_____

    e. Do all parts of humans grow at the same rate?

_____

_____

4. Parenting Skills

Animals with backbones are called *vertebrates*. They develop from fertilized eggs. Their ways of reproducing, however, show many differences. In fact, you could sort them into groups by answering a few questions about their fertilized eggs and what happens to them. Look at the chart on page 90 of your textbook.

    a. The letters *A* through *F* in the diagram represent headings that could be in the form of questions. For each letter, supply the question that is answered by that section of the chart. For example, the question for *A* could be, Where does fertilization take place?

_____

_____

_____

_____

_____

_____

_____

SCIENCEPLUS • LEVEL GREEN   27

Chapter 5 Review Worksheet, continued

b. Based on the information in the chart, which of the animals named in the far right column do you think would have the best chance to survive and to grow up? Which would be the least likely to survive? Give reasons for your choices.

_____

_____

_____

_____

_____

_____

_____

c. Why do you think a frog produces many more eggs at one time than a bird does?

_____

_____

_____

_____

_____

_____

_____

d. Which do you think produces more eggs at one time, a fish or a turtle? Why?

_____

_____

_____

_____

_____

_____

_____

Name _____ Date _____ Class _____

Chapter 5 Assessment

Word Usage

1. Show that you understand the following terms by using them in one or more sentences:

    a. regeneration, cancerous growth, starfish, tumor, renewal, continual growth, skin, hair

    _____
    _____
    _____
    _____
    _____

    b. head, proportion, size, adult, birth

    _____
    _____
    _____
    _____
    _____

Correction/ Completion

2. The statements below are either incorrect or incomplete. Make the statements correct and complete.

    a. The cells of a person's skin may start growing uncontrollably after the person has spent 3 hours a day in the sun for the past 15 years. This kind of growth is known as regeneration.

    _____
    _____
    _____

    b. The part of the body that grows the fastest from birth to the age of 5 is _____ .

SCIENCEPLUS • LEVEL GREEN 29

Chapter 5 Assessment, continued

Short Response

3. How is regeneration different from reproduction?

_____

_____

_____

_____

_____

_____

_____

CHALLENGE 1

Graph for Interpretation

4. Use the following graph showing the heights of three generations of the Kapasi family to answer the questions that follow.

× = first generation
☆ = second generation
○ = third generation

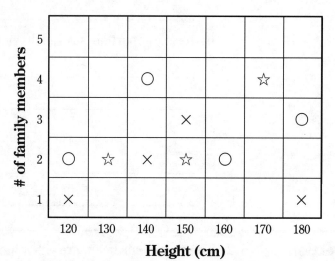

Height (cm)

a. How many members are in each of the first and second generations of the family? in the first and second generations combined?

_____

_____

_____

_____

30 UNIT 2 • PATTERNS OF LIVING THINGS

Chapter 5 Assessment, continued

b. What is the average height of each generation? of all three generations? Show your work.

c. Can you make any comments about height within a family based on your observations?

CHALLENGE 2
Short Essay

5. How might the continual growth of certain parts of an organism be beneficial? Mention at least one example in your answer.

Name _____ Date _____ Class _____

EXPLORATION 1

Chapter 6
Exploration Worksheet

Experimenting With Stimulus and Response: An Earthworm Responds, page 92

Cooperative Learning Activity		Safety Alert!
Group size	3 to 4 students	
Group goal	to test the responses of earthworms to stimuli	
Individual responsibility	Each member of your group should choose a role such as chief investigator, reporter, materials coordinator, or time-keeper.	
Individual accountability	Each group member should be able to explain how an earthworm senses stimuli such as touch, light, and odor.	

It is easy to test an earthworm's response to a number of stimuli. As you carry out these experiments, record your findings on the data sheet similar to the one on page 93 of your textbook.

The earthworm has no eyes, ears, or nose. Can it still sense light, sound, and odor, in addition to other stimuli?

You Will Need

- an earthworm
- a shallow pan
- paper towels
- water
- a pencil
- a flashlight
- vinegar

What to Do

Observe and record all responses in your ScienceLog using a data sheet like the one shown on page 93 of your textbook. Use the questions provided on the next page of this worksheet to guide you.

You are going to observe and investigate the earthworm's response to touch (pencil test), light (flashlight test), and smell (vinegar test). Clean up your area, and wash your hands with soap and water when you are finished.

1. Place the worm on damp paper towels in the shallow pan. (The worm will die if it is not kept moist.)

2. Gently touch (do not poke!) the side of the worm with the tip of a pencil. Then touch the worm on its back end.

3. Gently touch the worm on its front end several times.

4. Let the earthworm rest for a while under a damp paper towel. Remove the towel, and shine a beam of light on the front end of the worm. Give the worm another rest, and shine the light on other parts.

5. Soak a piece of paper towel in vinegar. Uncover the worm, and bring the towel near it. **Do not touch the worm with the towel.**

Name _____ Date _____ Class _____

Exploration 1 Worksheet, continued

Questions

1. a. Are all parts of an earthworm equally sensitive to touch?
   _____

   b. If there is a difference, which parts respond more to touch than others?
   _____
   _____

2. a. Which part of the worm seems most sensitive to light?
   _____

   b. Did the worm move toward the light (positive response) or away from the light (negative response)?
   _____

3. Give two reasons why earthworms are not usually found on the surface of the ground.
   _____
   _____
   _____

4. When are you likely to find earthworms on the surface? Why are they there at that time?
   _____
   _____
   _____
   _____

5. How could the earthworm sense the vinegar without touching it?
   _____
   _____

6. How are the earthworm's responses useful for its way of life?
   _____
   _____
   _____

SCIENCEPLUS • LEVEL GREEN   33

Name _____ Date _____ Class _____

**Chapter 6
Discrepant Event Worksheet**

Temperature in Degrees Cricket   Teacher Demonstration

Use this demonstration to introduce Lesson 3, which begins on page 100 of the textbook.

You Will Need

- 2 crickets (available in many bait stores)
- 2 jars
- 2 nylon stockings
- 2 rubber bands
- a watch with a second hand
- a thermometer

Advance Preparation

You may wish to try this activity in advance. First add one cricket to each jar. Cover each jar with a nylon stocking and secure with a rubber band. (Make sure that the nylon stocking is stretched sufficiently to allow enough oxygen into the jars.) Leave one jar at room temperature in the classroom and the other in a refrigerator long enough to slow the chirping of the cricket significantly (but not so long that the cricket is harmed!). Note: Be prepared to conduct this activity soon after you remove the jar from the refrigerator. When doing so, record the temperature of the refrigerator in degrees Fahrenheit.

What to Do

1. Have students observe the cricket that has been kept in the refrigerated jar. Tell students to count the number of chirps the cricket makes in 15 seconds and then add 40 to that number. Record both of these numbers in a data chart on the chalkboard.

2. Record the temperature of the refrigerator in degrees Fahrenheit in a separate column of the data chart. Point out the similarity between this number and the data from step 1—they should be roughly equal.

3. Have students repeat step 1 with the cricket that has been kept at room temperature.

4. Measure and record the temperature of the room in degrees Fahrenheit. Again, the temperature of the room and the data from step 3 should be almost equal.

What's Happening Here?

Encourage class discussion, using the following questions as a guide:

1. What general conclusions can you make about your observations? *(Student answers should reflect the relationship between temperature and chirping frequency. Students may also conclude that a cricket can serve as a kind of thermometer.)*

2. If we refer to the number of chirps in 15 seconds as "degrees cricket," what is the corresponding temperature in degrees cricket for 82°F? *(82 − 40 = 42 degrees cricket)* At what temperature would a cricket stop chirping? *(40 − 0 = 40°F)* What might this suggest about the cricket? *(This might suggest that the cricket is in a state of dormancy due to cold weather.)*

34   UNIT 2 • PATTERNS OF LIVING THINGS

Name _____ Date _____ Class _____

EXPLORATION 3

Chapter 6
Exploration Worksheet

Which Loses Heat Faster—A Mouse or a Mountain Lion? page 103

| Your goal | to decide if the size of an organism affects the rate of heat loss | Safety Alert! |

You Will Need

- 2 glass bottles (with lids), one of which can hold twice as much as the other
- 2 alcohol thermometers
- a pitcher
- hot water from the tap
- oven mitts

What to Do

1. Fill each bottle with hot water at the same time.
2. Record the temperature of the water in each bottle. Put on the lids.
3. Measure the temperature of the water in each bottle every 5 minutes for 30 minutes. **Caution: Be careful not to burn yourself with the hot water. Use oven mitts as neccessary to handle the bottles of hot water.**

Was there any difference in the rate of cooling of the small and large containers? Try to explain the difference.

_____

_____

_____

_____

_____

_____

Time (min.)	Temp. of small bottle	Temp. of large bottle
5		
10		
15	*Example*	
20		
25		
30		

Does the result support your prediction about the mouse and the mountain lion?

_____

_____

_____

_____

_____

_____

SCIENCEPLUS • LEVEL GREEN 35

Name _____ Date _____ Class _____

Chapter 6 Activity Worksheet

A Warmblooded Puzzle

Complete this activity after finishing Lesson 3, which begins on page 104 of your textbook.

For each item in list I, select the best matching item in list II. (You may have to do a little research on a few of the items.) Enter the letter in the appropriately numbered box in the puzzle square. One letter is used more than once. When you have completely filled the boxes, you will discover a "warmblooded" message!

List I
1. warmblooded winter inhabitant
2. migrates south in winter
3. dormancy
4. panting
5. feather fluffing
6. sweating
7. coldblooded
8. banding
9. rabbit ears

List II
A. robin
E. method of tracking migrating birds
H. slowed breathing and heartbeat
O. way of cooling off
R. way of keeping warm
T. Blood gives off heat through these.
U. Body temperature matches that of surroundings.
Y. black-capped chickadee

1	2	3
4	5	6
7	8	9

36 UNIT 2 • PATTERNS OF LIVING THINGS

Name _____ Date _____ Class _____

Chapter 6
Theme Worksheet

Survival or Extinction?   Teacher's Notes

This worksheet is designed to complement the theme connection outlined on page 108 of the Annotated Teacher's Edition. It is also designed as an extension of Plants and Animals Made to Order, on page 109 of the Pupil's Edition.

Focus question	If a species from an equatorial region were to move to one of the coldest regions of the Earth, what new adaptations would help it to survive in this new habitat?

Advance Preparation

You may create the playing pieces for this game in advance, or you may have students create the pieces. First prepare several cups with the following labels: Skin covering, Diet, Special adaptations, and Climate. Then, on slips of paper, write each entry from the Playing Pieces table on the next page. Place these playing pieces in the corresponding cups. Divide the class into groups of three to four students.

Explaining the Game

An animal's chances for survival are determined in part by how well the adaptations of its species are suited to a given climate. In this game, the survival of each animal will be determined by a rating system. The rating system consists of four categories, which are listed on the next page. Certain adaptations fare better in particular climates. To evaluate an adaptation in a given climate, the following notation is used: A = Average; E = Excellent; P = Poor.

Playing the Game

1. One student draws a Climate piece to establish the environment. Each student then draws one piece from each of the other three cups in order to "create" an animal that has those three adaptations.

2. To create the most successful animal for survival in a given climate, each student may do *one* of the following for each of the three adaptation pieces they draw:
   • Trade for another group member's piece (must be the same adaptation type).
   • Replace one piece with another piece from the same cup.
   • Keep the piece that they originally drew from the cup.

3. Once all trades and discards have been completed, students should reveal their finished animals and determine their overall ratings based on the tables. The animal with the highest rating wins that round. If an animal has a rating of poor for each adaptation, it becomes extinct.

4. Students can continue to play the game as long as time allows, creating a new animal in every round. There should be enough rounds of play for each student to have a chance to create the most successful animal in a given environment.

SCIENCEPLUS • LEVEL GREEN 37

Name _____ Date _____ Class _____

Survival or Extinction?

This game corresponds to page 109 of your textbook.

Chapter 6 Theme Worksheet

What to Do

To learn more about adaptations and survival, use the following charts as directed by your teacher:

Playing Pieces

Skin covering
light fur
heavy fur
scales
bare damp skin
bare dry skin
thick hide
exoskeleton
feathers

Diet
eats nuts and berries
eats insects
eats grasses
eats fish/shellfish
eats leaves
eats big animals
eats small animals

Special adaptations
hibernates
migrates
lives in trees
drinks very little
white color
tan color
green color
brown color

Climate
desert
jungle
forest
polar

Rating System

	Climate			
Skin covering	Desert	Jungle	Forest	Polar
light fur	A	A	E	P
heavy fur	P	P	E	E
scales	E	E	A	P
bare damp skin	P	E	A	P
bare dry skin	E	P	A	P
thick hide	A	A	A	A
exoskeleton	E	E	A	P
feathers	A	A	A	E
Diet				
eats nuts and berries	P	E	E	A
eats insects	E	E	A	P
eats grasses	P	P	A	A
eats fish/shellfish	P	E	A	E
eats leaves	P	A	E	P
eats big animals	A	A	A	A
eats small animals	E	A	A	A
Special adaptations				
hibernates	A	P	E	E
migrates	A	P	A	E
lives in trees	P	E	E	P
drinks very little	E	P	A	A
white color	P	P	P	E
tan color	E	P	A	P
green color	P	E	A	P
brown color	A	A	E	P

38 UNIT 2 • PATTERNS OF LIVING THINGS

Name _____ Date _____ Class _____

Chapter 6 Review Worksheet

Challenge Your Thinking, page 110

1. An Animal's Tale (Tail?)

Each member of the class should choose a specific animal and tell a story, from that animal's point of view, about its preparations for winter. If you choose to be an animal that migrates, tell the class how, when, and where you migrate and the reasons for your migration. If you choose to be an animal that stays put, describe your preparations. Record your story in your ScienceLog.

You will have to do research. Look in books on the subject, and ask informed people. Try to make your story about 150 words long.

Illustration also on page 110 of your textbook

2. I Can't Say No to Sweets

In your ScienceLog, describe how you would test the responses of ants to the stimuli of sugar, moisture, and heat. Use labeled diagrams. Make a prediction about what the ants' responses will be.

SCIENCEPLUS • LEVEL GREEN   39

Name _____ Date _____ Class _____

Chapter 6 Review Worksheet, continued

3. Good Reflexes

Look at the picture below. Pick out all of the examples of stimuli and responses to these stimuli that you can find. Determine whether each response is positive (+) or negative (−), and record your answers in the following chart:

Stimulus	Response	+	−

Illustration also on page 111 of your textbook

40 UNIT 2 • PATTERNS OF LIVING THINGS

Name _____ Date _____ Class _____

Chapter 6 Review Worksheet, continued

4. **Strange Behavior**

Read the conversation below. Write at least two possible explanations for each animal behavior mentioned.

Pamela: I see ants going up and down the stems of our rosebush. Why do they do that?

_____
_____
_____

Mike: I have a mystery too. A lot of crows fly into the trees behind our apartment building late in the afternoon each day.

_____
_____
_____

Ani: Some kind of animal dug up big patches of our lawn last night.

_____
_____
_____

Raoul: There's a bird that pecks at one of our windows early every morning. Why does it do that?

_____
_____
_____

Name three things you can do to improve your grade for the next marking period?

_____
_____
_____
_____
_____

Name _____ Date _____ Class _____

Chapter 6
Assessment

Word Usage

1. Show that you understand the following terms by using them in one or two sentences:

   a. descendants, distance, humpback whales, migrate, return

   _____

   _____

   _____

   b. clocks, daylight, schedule, animals

   _____

   _____

   _____

Short Response

2. Explain how negative responses help an organism.

   _____

   _____

   _____

   _____

   _____

   _____

Correction/ Completion

3. The statements below are either incorrect or incomplete. Make the statements correct and complete.

   a. Animals who live in arctic climates have long, thin limbs in order to keep themselves warm.

   _____

   _____

   _____

   _____

42 UNIT 2 • PATTERNS OF LIVING THINGS

Chapter 6 Assessment, continued

b. Structural adaptations in species develop rapidly so that animals can respond to long-term changes in their environments.

___

CHALLENGE 1

Answering by Illustration

4. Imagine a swampy, tropical habitat where annual temperatures range from 25°C to 30°C. The area receives about 250 cm of rainfall each year and is home to many plants and animals, including fish, frogs, snakes, rodents, birds, and alligators.

   Design your own warmblooded, meat-eating animal to live in this environment. On a separate sheet of paper, draw a picture of your animal. Then add labels to indicate what structures the animal has for locomotion, catching prey, protecting itself from predators, and protecting itself from harmful elements in the environment (such as direct sunlight).

CHALLENGE 2

Short Essay

5. Consider a shark and a dog. Describe the different ways they respond to their individual environments over the course of a year.

Name _____ Date _____ Class _____

Chapter 7 Resource Worksheet

How to Use a Microscope, page 114

The instrument that allows us to see cells and examine small objects is the microscope. As you read the instructions for its use on page 114 of your textbook, try to match the names of the parts of the microscope (italicized below) with the labels on the diagram. The diagram below shows only one example of a microscope. The microscope that you use in your own classroom may be slightly different. For example, it may have a mirror instead of its own light source.

light source, base, eyepiece, tube, stage, diaphragm, objective lenses, nosepiece, coarse-adjustment knob, fine-adjustment knob

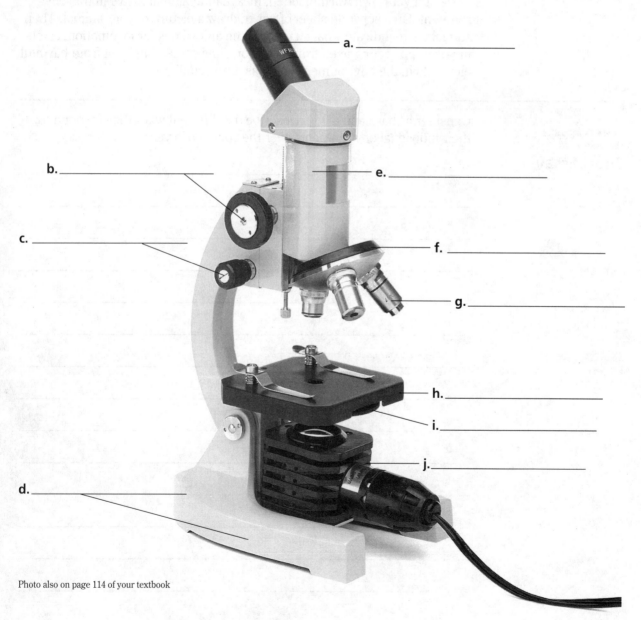

a. _____
b. _____
c. _____
d. _____
e. _____
f. _____
g. _____
h. _____
i. _____
j. _____

Photo also on page 114 of your textbook

Name _____ Date _____ Class _____

Chapter 7
Transparency Worksheet

Compound Light Microscope   Teacher's Notes

This worksheet corresponds to Transparency 17 in the Teaching Transparencies binder.

Suggested Uses

Use as a visual aid with the topic listed below.

How to Use a Microscope, page 114

Use in connection with the following activities:

a. Call students' attention to the labels on the microscope. Explain to them the importance of knowing each part as well as its function. (Note: This transparency shows certain parts of the microscope that are not featured on page 114 of the textbook, thus allowing you to expand the students' knowledge of the microscope.)

b. Have students work together and quiz each other on the functions of each part of the microscope.

c. Use the transparency as a reference during class activities that involve the microscope.

Use the transparency with the transparency worksheet on the next page for reteaching or review. Please note: The worksheet is a reproduction of the actual transparency with certain labels omitted. For answers to that worksheet, see the transparency.

Possible Extension Question

What is the proper method for carrying a microscope?

Answer to Extension Question

The microscope should be carried with one hand holding the microscope by its arm and the other hand supporting the base of the microscope.

SCIENCEPLUS • LEVEL GREEN   45

Name _____ Date _____ Class _____

**Chapter 7
Transparency Worksheet**

Compound Light Microscope

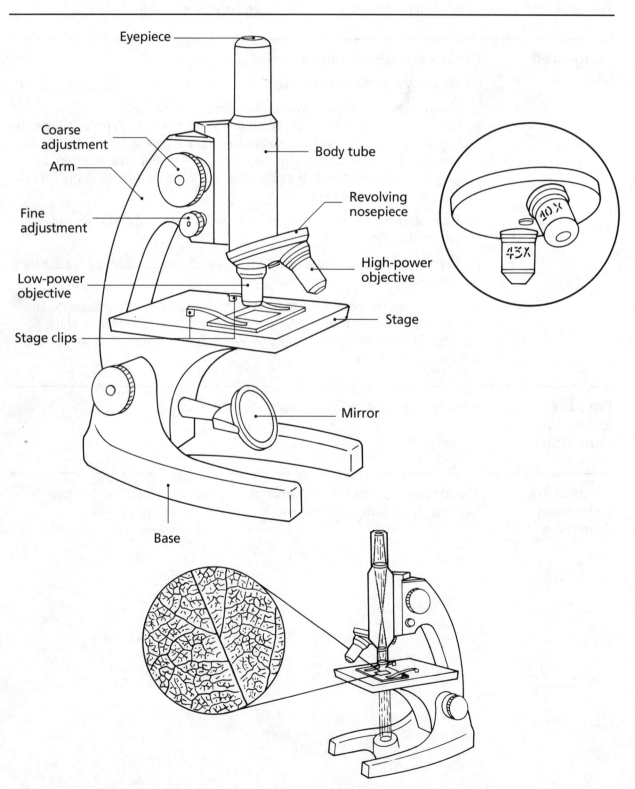

UNIT 2 • PATTERNS OF LIVING THINGS

Name _____ Date _____ Class _____

EXPLORATION 1

Chapter 7
Exploration Worksheet

Looking at Cells, page 115

| Your goal | to prepare a wet mount for a microscope |

When you examine the cells of a plant or animal through a microscope, the sample you use must be very thin so that light can shine through it. Also, the sample must be kept moist. The illustrations on page 115 of your textbook show how to make a **wet mount**.

You Will Need

- a microscope
- microscope slide(s)
- coverslip(s)
- an eyedropper
- a small piece of newspaper with a letter *e*
- forceps
- water
- tissue

What to Do

In this activity, you will see that things look different, not just larger, under a microscope.

Making a Wet Mount

1. Put a drop of water on a slide.
2. Place the piece of newspaper with the letter *e* on the water.
3. Lean a coverslip against a pencil. Slowly lower the pencil, and finally remove it. Avoid getting air bubbles under the coverslip. Absorb excess moisture around the coverslip with tissue.
4. Now view the slide through the microscope, first with the low-power lens and then with the high-power lens. In your ScienceLog, draw what you see.

Analyze Your Work

1. If you move the slide to the left, in which direction do the objects in the field of view move?

   _____

2. When you change from the low-power lens to the high-power lens, how does your field of view change?

   _____

Watch Out for These!

Dark, round circles are air bubbles.

Dark, jagged lines are coverslip edges.

Thin, irregular lines are caused by water drying up.

> To calculate total magnification, multiply the magnification of the eyepiece lens by the magnification of the objective lens you are using. A 10× eyepiece lens and a 10× objective lens will give a total magnification of 100×.

SCIENCEPLUS • LEVEL GREEN 47

Name _____ Date _____ Class _____

Chapter 7
Transparency Worksheet

Plant and Animal Cells   Teacher's Notes

This worksheet corresponds to Transparency 20 in the Teaching Transparencies binder.

Suggested Uses

Use as a visual aid with any of the topics listed below.

Skilled Observations of Cells, page 116
Unit 6, Energy and You
Unit 7, Temperature and Heat

Use in connection with the following:

a. Extend the study of cell structure in Lesson 2, Skilled Observations of Cells. Explain to students the difference between plant and animal cells.

b. As students explore cheek cells and onion-skin cells, have them label any parts of the cells they see while doing Exploration 2, on page 116 of the textbook.

c. Extend the concept of energy in cells in Unit 6 or the concept of converting heat energy in Unit 7.

Use the transparency with the transparency worksheet on the next page for reteaching or review. Please note: The worksheet is a reproduction of the actual transparency with certain labels omitted. For answers to that worksheet, see the transparency.

Possible Extension Questions

1. What are the basic differences between a plant cell and an animal cell?
2. What is the function of the cell membrane?
3. What is the function of the mitochondria in animal and plant cells?
4. What is the function of the chloroplasts in plant cells?

Answers to Extension Questions

1. The plant cell has a cell wall and chloroplasts, but the animal cell does not.
2. The function of the cell membrane is to control what materials enter and leave the cell.
3. The function of the mitochondria is to release energy from food molecules.
4. The function of the chloroplasts in plant cells is to convert light energy into chemical energy.

Name _____ Date _____ Class _____

Chapter 7
Transparency Worksheet

Plant and Animal Cells

1. _____
2. _____
3. _____
4. _____
5. _____
6. _____
7. _____
8. _____

9. _____
10. _____
11. _____
12. _____

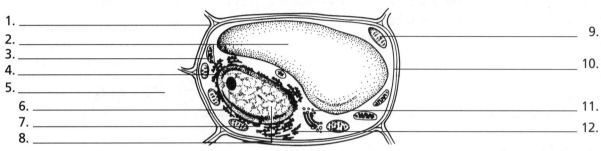

Plant Cell

13. _____
14. _____
15. _____
16. _____
17. _____
18. _____
19. _____

20. _____
21. _____
22. _____
23. _____

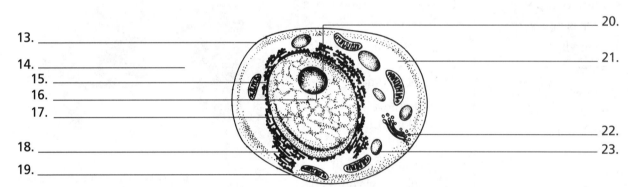

Animal Cell

24. _____
25. _____

26. _____

27. _____

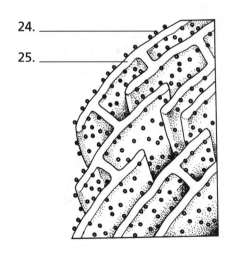
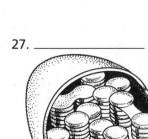
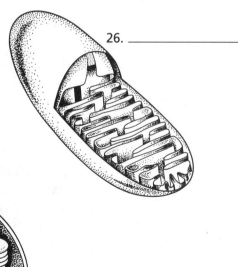

SCIENCEPLUS • LEVEL GREEN 49

Name _____ Date _____ Class _____

EXPLORATION 2

Chapter 7 Exploration Worksheet

What Are Plants and Animals Made Of? page 116

| **Your goal** | to observe, draw, and distinguish between plant and animal cells | **Safety Alert!** Iodine is an eye irritant and is somewhat corrosive to the skin. Exercise caution while using it. |

You Will Need

- a microscope
- microscope slides
- coverslips
- water
- a knife
- eyedroppers
- one-quarter of an onion
- forceps
- iodine solution
- a prepared slide of human cheek cells

What to Do

1. Make a wet mount of a piece of onion skin.

   a. Remove the outside layer of one-quarter of an onion.

   b. Remove the thin skin from the inside of the layer.

   c. Cut off a small piece for your slide.

   d. Place the onion skin on a slide, add a drop of water, and place a coverslip on top.

2. Adjust the diaphragm and the light source to get the best lighting. Observe the onion skin under low power and then high power.

3. Prepare a second wet mount, but this time add a drop of iodine solution to the onion skin. How does this change what you see?

   _____
   _____

4. As you look at the onion skin, do you observe a pattern of very small parts that all look very much alike? These are the cells. In your ScienceLog, draw two or three onion-skin cells under low power and then under high power. Make note of the magnification used each time.

Illustration also on page 116 of your textbook

50 UNIT 2 • PATTERNS OF LIVING THINGS

Name _____ Date _____ Class _____

Exploration 2 Worksheet, continued

5. In your ScienceLog, draw all of the features that you see in one cell. Label the cell and the **cell wall** (the thick casing of the cell).

6. Clean and dry the slides and coverslips.

7. Look at the prepared slide of material taken from the inside of a person's cheek. Notice the color due to the stain used. Instead of iodine, other stains are often used.

8. What do you observe through the microscope? In your ScienceLog, make drawings under low and high power as you did for the onion material. Don't forget to note the magnification used. Show all detailed features as accurately as possible.

What's the Difference?

You have seen some plant cells—in the skin of an onion. You have also seen some animal cells—human cheek cells. Consult your drawings of cells, and use your memory to answer the following questions:

1. What characteristics do plant and animal cells share?

   _____
   _____
   _____
   _____
   _____
   _____

2. Would you be able to distinguish onion-skin cells from human cheek cells? If so, how would you do it? Record your thoughts here and think about this question as you carry out the next Exploration.

   _____
   _____
   _____
   _____
   _____
   _____
   _____

SCIENCEPLUS • LEVEL GREEN

Name _____ Date _____ Class _____

EXPLORATION 3

**Chapter 7
Exploration Worksheet**

Edible Cells, page 118

| Your goal | to observe, draw, and distinguish between plant and animal cells | Safety Alert!  |

Part 1: Plant Cells

You Will Need

- a microscope
- microscope slides
- coverslips
- a knife
- 2 straight pins
- iodine solution
- water
- eyedroppers
- pieces of plant material: potato, lettuce, green pepper, banana peel, orange peel, and carrot (optional: tea, nutmeg, pepper, mustard, coffee, and ginger)

What to Do

1. Cut or scrape off sections as thin as possible from the potato and from the banana and orange peels. Try to include a piece of skin along with scrapings of the flesh of the potato. Use the straight pins to pull the material apart. If your specimens are too thick, you may find that high power does not show details well.

2. Mount each type of plant on two slides. Use only water to mount one slide, and use a drop of iodine as stain on the other.

3. In your ScienceLog, draw the cells you observe, and label each drawing.

4. Clean up your lab area and wash your hands after you are finished.

Questions

1. Did you observe any round objects (other than air bubbles)? What do you think these might be? Of what use might they be to the plants?

   _____

   _____

2. When the iodine was used, did you notice a color on some slides that didn't appear on the other slides? What was the color?

   _____

3. Iodine is used as a test for the presence of starch, one of the substances in foods. Adding iodine to starch results in a blue to purple-black color. Some of the substances you examined do contain starch particles. Which plants observed gave a positive test for starch?

   _____

4. Can you think of other sources of starch in your diet?

   _____

   _____

52   UNIT 2 • PATTERNS OF LIVING THINGS

Name _____ Date _____ Class _____

Exploration 3 Worksheet, continued

Part 2: Animal Cells

You Will Need

- a tiny piece of beef liver
- the same equipment used in Part 1

What to Do

1. Put a tiny piece of liver on a slide.
2. Pull apart the piece of liver using two straight pins so that the material is as thin as possible.
3. Make a wet mount. Carefully examine the sample under low and high power.
4. Add a drop of iodine to a second piece of liver as a stain, and observe this slide under high and low power as well.
5. In your ScienceLog, make two drawings of a liver cell, one with iodine and one without. Show as much detail as you can.

Questions

1. Now that you have observed cells closely, write a two-sentence answer to the question, "What are cells like?"

   _____
   _____
   _____

2. In Exploration 2, you compared plant and animal cells. Write any further observations you have that would help you distinguish between plant and animal cells.

   _____
   _____
   _____
   _____

Plant or Animal?

Look at the photos on page 119 of your textbook. Sort them into plant or plant-like cells and animal or animal-like cells. In your ScienceLog, state your reasons for your decisions. (Hint: Pay attention to the outer boundary of each cell. Does it appear rigid, thick, and definite in shape or delicate and flexible?)

Plant or plant-like	Animal or animal-like

Name _____ Date _____ Class _____

Chapter 7
Review Worksheet

Challenge Your Thinking, page 120

1. A Cellular Questionnaire

Quiz yourself with the following questions:

a. How are the cells that make up living things like bricks?

_____
_____
_____

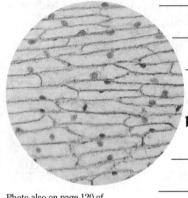

Photo also on page 120 of your textbook

b. How are cells not like bricks?

_____
_____
_____
_____

c. How do microscopes aid in the study of living things?

_____
_____
_____
_____

d. Describe how things look different under a microscope (other than just looking larger).

_____
_____
_____
_____

e. Suppose that you are looking through a microscope at a tiny animal. It swims up and to the right and moves out of your field of view. Which way do you move the slide to follow it?

_____

54 UNIT 2 • PATTERNS OF LIVING THINGS

Name _____ Date _____ Class _____

Chapter 7 Review Worksheet, continued

2. Microscope Champions

Use the table below to rate yourself as a microscope user.

Activity	Good	Fair	Could be better
a. I carry the microscope carefully to my work area.			
b. I begin observing with the lowest power objective lens.			
c. I carefully adjust the light level.			
d. I avoid touching the objective lens to the slide I am viewing.			
e. I focus the microscope by first using the coarse-adjustment knob and then using the fine-adjustment knob.			
f. I properly prepare the microscope for storage when my work is finished.			

3. Look and See

Examine the photograph of the cell below.

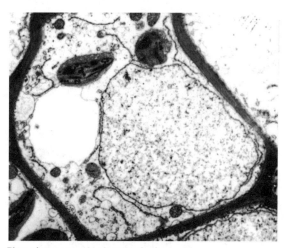

Photo also on page 121 of your textbook

Is it a plant or an animal cell? How can you tell?

_____

_____

Use the following words to label your drawing: *cell wall, vacuole, chloroplast,* and *nucleus.* Where would the cell membrane be located if it were visible? Label this also.

SCIENCEPLUS • LEVEL GREEN   55

Name _____ Date _____ Class _____

Chapter 7 Review Worksheet, continued

4. **Mystery Word** Fill in the correct "cellular" words below to find out what this unit is all about.

☐☐☐☐☐☐☐☐ living matter found outside the nucleus of the cell

☐☐☐☐☐☐ space used for storage (larger and more common in plant than in animal cells)

☐☐☐☐☐☐☐☐☐☐☐ green, often oval, structure in plant cells; used in food production

☐☐☐☐☐☐☐☐☐☐☐☐ oval bodies containing folded membrane; sites of cell activity where energy is released

☐☐☐☐ ☐☐☐☐ thick outer covering of the cell; provides strength and protection

☐☐☐☐ ☐☐☐☐☐☐☐☐ outer edge of an animal's cell

☐☐☐☐☐☐☐ dense, round or oval body; controls the life activities of the cell

☐☐☐☐☐ building blocks of all living things

56 UNIT 2 • PATTERNS OF LIVING THINGS

Name _____ Date _____ Class _____

Chapter 7 Assessment

Word Usage

1. For each group of words below, write one or more sentences. Use all of the words in the group to show how they are related.

    a. nucleus, support, cell wall, separates, activities

    _____
    _____
    _____

    b. stage clips, slide, low-power lens, coarse-adjustment knob

    _____
    _____
    _____
    _____

Correction/Completion

2. The statements below are incorrect or incomplete. Your challenge is to make them correct and complete.

    a. The part of a microscope that controls the amount of light passing through an object is the eyepiece.

    _____
    _____

    b. All cells have both a cell membrane and a cell wall.

    _____
    _____

Numerical Problem

3. Tami is using a microscope that has the powers listed below.

Eyepiece = 10×	Low-power lens = 4×
Medium-power lens = 10×	High-power lens = 46×

    How many times larger will objects appear if she uses the low-power lens? the high-power lens?

    _____
    _____

SCIENCEPLUS • LEVEL GREEN 57

Name _____ Date _____ Class _____

Chapter 7 Assessment, continued

Illustration for Interpretation

4. Below is a good drawing of a piece of onion skin viewed under a microscope. Why is it a good drawing? List at least four reasons.

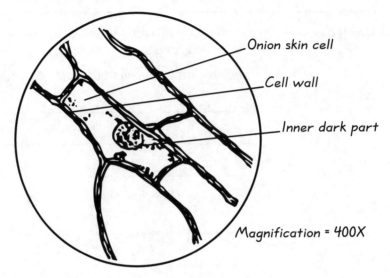

Onion Skin Cell

_____
_____
_____
_____
_____
_____
_____

Short Response

5. Why don't animal cells have cell walls?

_____
_____
_____
_____
_____
_____

58 UNIT 2 • PATTERNS OF LIVING THINGS

A Magic Square

Complete this activity as you conclude Unit 2.

Match the items in list I with their descriptions in list II. Enter the number of the description in the box of the magic square containing the letter of the appropriate item. Add the horizontal rows, vertical rows, and diagonals to discover the MAGIC.

List I
A. snail
B. variable
C. tree rings
D. adaptation
E. positive response
F. regeneration
G. earthworm
H. butterfly
I. cancer
J. cells
K. dormancy
L. migration
M. germination
N. polar bear
O. spider
P. stimulus

List II
2. Growing roots, stems, and leaves from a seed
3. Moving to a new location in the fall or spring
4. An indication of age
5. A crab growing a new claw
6. Period when the activities of life slow down
7. A warmblooded animal
8. Motion toward light or touch
9. A change in a living thing that occurs as a result of a change in the environment
10. A condition that is changed in an experiment
11. Has segments with bristles
12. Touching the leaflets of a mimosa plant
13. Uncontrolled, harmful growth
15. An animal with six legs
16. An animal with one foot
17. Small "bricks" of living things
18. An animal with eight legs

A	B	C	D
E	F	G	H
I	J	K	L
M	N	O	P

Magic number: _____

Name _____ Date _____ Class _____

Unit 2
Unit Review Worksheet

Making Connections, page 122

The Big Ideas In your ScienceLog, write a summary of this unit, using the following questions as a guide:

1. What are some signs of life? (Ch. 4)
2. How do plant and animal movement differ? (Ch. 4)
3. In what ways are the growth patterns of plants and people similar? (Ch. 5)
4. In what ways are they different? (Ch. 5)
5. What are some of the different forms of growth? (Ch. 5)
6. What is the importance of each? (Ch. 5)
7. What does "response to stimuli" mean? (Ch. 6)
8. Why is this response valuable? (Ch. 6)
9. What evidence is there that animals have a sense of time? (Ch. 6)
10. What are some adaptations of plants and animals over long periods of time? (Ch. 6)
11. How do plant and animal species adapt over long periods of time? (Ch. 6)
12. How are plant and animal cells similar? How do they differ? (Ch. 7)
13. Describe the procedure for making a wet mount. (Ch. 7)

Checking Your Understanding

1. Would a seed be an example of a living thing? Why or why not?

   _____
   _____
   _____

2. What are some of the reasons that animals migrate? Name at least three animals that migrate.

   _____
   _____
   _____
   _____

60 UNIT 2 • PATTERNS OF LIVING THINGS

Unit 2 Review Worksheet, continued

3. Look at the illustration below showing the life cycle of a butterfly. How does the butterfly's growth pattern differ from that of a human?

   _____

   _____

   _____

   _____

Illustration also on page 123 of your textbook

4. *concept map* Construct a concept map using the following terms: *ectothermic, bear, endothermic, animals,* and *frog.*

Name _____ Date _____ Class _____

Unit 2 Review Worksheet, continued

5. What are some of the different ways in which animals move?

   _____

   _____

6. Are all forms of growth useful? Explain.

   _____

   _____

   _____

7. Is bigger always better? Explain.

   _____

   _____

   _____

8. How do snakes demonstrate adaptation to their environment?

   _____

   _____

   _____

9. In your own words, compare an earthworm's locomotion to that of a snake.

   _____

   _____

   _____

   _____

10. How might a sense of time be helpful to animals? to plants?

    _____

    _____

    _____

    _____

    _____

Unit 2 Review Worksheet, continued

11. Add the following to your list of "Curious Questions." Suggest answers to each.

    a. Where do snakes, frogs, and salamanders go in the winter?

    _____
    _____
    _____

    b. How do endothermic animals keep their body temperature constant?

    _____
    _____
    _____
    _____
    _____

    c. What advantages do endothermic animals have over ectothermic animals?

    _____
    _____
    _____
    _____
    _____

    d. What causes animals to start migrating? to stop migrating?

    _____
    _____
    _____
    _____

    e. What cues do animals use to guide them as they migrate?

    _____
    _____
    _____
    _____

Name _____ Date _____ Class _____

Unit 2
End-of-Unit Assessment

Word Usage

1. For each group of words below, write one or two sentences. Use all of the words in the group to show how they are related.

   a. stimulus, respond, worm, plant, light

   _____

   _____

   _____

   b. cooling system, sweating, panting, dog

   _____

   _____

   _____

Correction/ Completion

2. The statements below are either incorrect or incomplete. Make the statements correct and complete.

   a. When a tree is cut, growth rings may be visible. The number of rings tells the tree's _____, and the width of a ring shows how much the tree _____ that year.

   b. After a lobster loses a claw in a fight, the claw may _____. This is an example of regeneration.

   c. Humpback whales are dormant in winter; they travel south to warmer waters.

   _____

   _____

Short Response

3. Tell how a cactus plant is adapted to its environment.

   _____

   _____

   _____

   _____

64   UNIT 2 • PATTERNS OF LIVING THINGS

Unit 2 Assessment, continued

Graph for Correction or Completion

4. A bean plant is watered regularly for 4 weeks. After that, it is no longer watered. Show what the graph might look like after the fourth week.

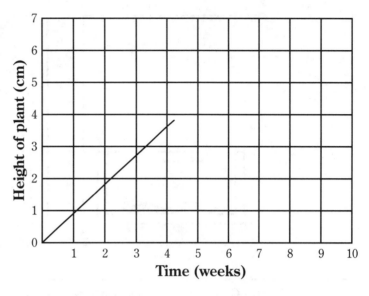

Illustrations for Correction or Completion

5. Complete the drawing by showing how the plant in the illustration would look 3 weeks after being moved away from the light.

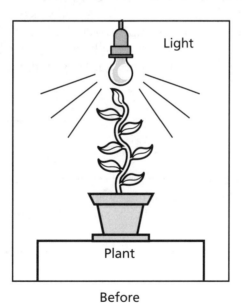
Before

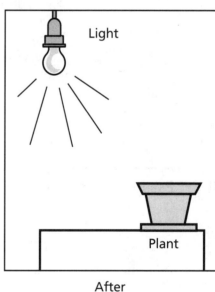
After

Name _____ Date _____ Class _____

Unit 2 Assessment, continued

6. Complete the drawing by showing the location of the worm in the box after more light has been allowed into the box.

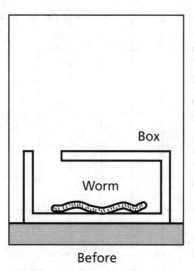
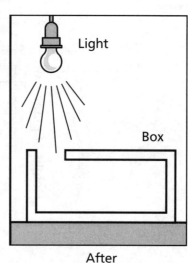

Before   After

CHALLENGE 1

Numerical Problem

7. **The Great Worm Race Between Alpha, Beta, and Gamma**
The three worms started from the center of the ring. Alpha sped outward at 50 cm/h for the first half-hour of the one-hour race. During the last half-hour, his speed was only 30 cm/h. Beta, on the other hand, started out slowly; during the first half-hour her speed was only 20 cm/h. However, in the final half-hour, she sped up to 70 cm/h. Gamma had a steady pace for the full hour, averaging 40 cm/h. Who went the farthest in 1 hour and won the race? How far did the winner go? How far did the losers go? Show your work.

_____
_____

Short Essay

8. Describe an imaginary, migratory animal. Identify the path of its migration, when it migrates, and what factors cause it to migrate.

_____
_____
_____
_____
_____
_____
_____
_____

Name _____ Date _____ Class _____

Unit 2 Assessment, continued

Illustration for Interpretation

9. Study the illustration below showing the life cycle of a frog.

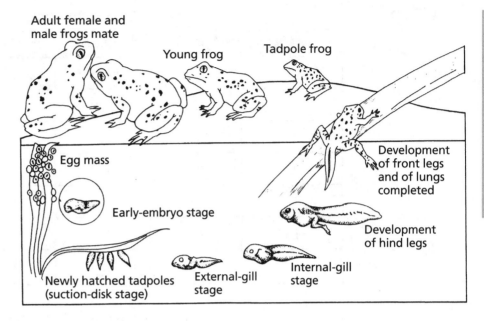

a. Which features of the tadpole are adaptations to life in the water?

_____

_____

b. How does the tadpole move?

_____

_____

c. Which features of the frog are adaptations to life on land?

_____

_____

d. How does the frog move on land?

_____

_____

e. Why is water important to the frog's reproduction?

_____

_____

_____

SCIENCEPLUS • LEVEL GREEN 67

Name _____ Date _____ Class _____

Unit 2 Assessment, continued

Data for Interpretation

10. Consider this table of students' heights.

Average Height (cm)

Grade	Boys	Girls	Raj	Marilyn
6	125	125	115	138
7	128	130	118	153
8	132	136	122	166
9	138	143	126	170
10	148	149	135	171
11	161	154	155	172
12	172	158	173	172

_____ a. During which grade did the average height for girls increase the most?

_____ b. During which grade did the average height for boys increase the most?

_____ c. During which grade did Raj grow the most?

_____ d. During which grade did Marilyn grow the most?

e. Using the data above for boys and girls, complete the bar graph below.

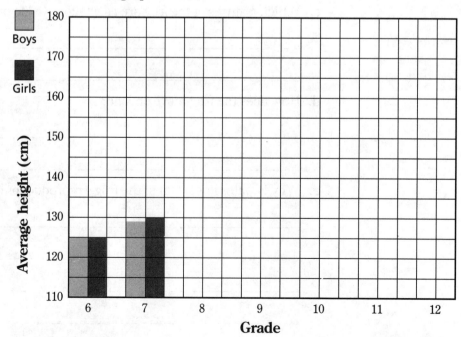

68   UNIT 2 • PATTERNS OF LIVING THINGS

Name _____ Date _____ Class _____

**Unit 2
Activity Assessment**

What's in the Water?   Teacher's Notes

Overview	Students analyze samples of pond water for living organisms, decaying materials, and nonliving components. They make observations, prepare slides, use a microscope, and compile a list of living, once-living, and nonliving things.	Safety Alert!

Materials
(per activity station)

- a magnifying glass
- an eyedropper
- a microscope
- a pair of tweezers
- microscope slides
- coverslips
- paper towels
- pond-water sample (approximately 0.5 L)*

Ideally, the pond-water sample should include bits of soil or rock from the pond, plant material, insects or insect eggs, decaying organic material, and microscopic organisms.

Preparation

Prior to the assessment, equip student activity stations with the materials needed for each experiment. Remind students to use care when viewing the slides they make; light levels on the microscope should be kept relatively low for the protection of the living organisms.

Time Required

Each student should have 30 minutes at the activity station.

Performance

At the end of the assessment, students should turn in the following:
- a completed Data Chart
- a classification system for the living, once-living, and nonliving material found in the pond water

Evaluation

The following is a recommended breakdown for evaluation of this Activity Assessment:
- 25% appropriate use of materials and equipment
- 40% quality and thoroughness of observations
- 35% logical and complete classification system

SCIENCEPLUS • LEVEL GREEN   69

Name _____ Date _____ Class _____

Unit 2 Activity Assessment

What's in the Water?

At the station in front of you is a sample of water taken from a pond. Your challenge is to find as many living, nonliving, and once-living things as you can in the sample. Study the sample closely. Your observations should be made in three ways—with your eyes alone, with the magnifying glass, and with the microscope (to view the slides you will make). As you complete the following tasks, you'll be amazed to find out what's in the water!

Before You Begin. . . As you work through the tasks, keep in mind that your teacher will be observing the following:
- how you use the microscope
- how complete your observations are
- how logical and thorough your classification system is

Jump in!

Task 1 As you analyze your sample, complete the Data Chart on the next page. You may sketch or write down your observations. These should include, but are not limited to, the size and shape of the object, its color, whether it moves and how it moves, and its relationship to other objects. Be as thorough and clear as you can when collecting your data.

Task 2 Now that you have made all of your observations, you must decide which of the things that you saw are living, once-living, and nonliving. Create a classification system on your own paper to organize the data on your chart. In your group of living things, indicate whether the organisms are plants or animals or whether you were unable to tell.

Name _____ Date _____ Class _____

Activity Assessment, continued

Data Chart

Material, substance, or organism	Observations

On a separate piece of paper, create a classification system summarizing the information in this chart.

SCIENCEPLUS • LEVEL GREEN

Name _____ Date _____ Class _____

Unit 2
Self-Evaluation

Self-Evaluation of Achievement

The statements below include some of the things that may be learned when studying this unit. If I have put a check mark beside a statement, that means I can do what it says.

_____ Given an example, I can tell whether a thing is living, once-living, or nonliving and explain why. (Ch. 4)

_____ I can explain at least one difference between animals and plants. (Ch. 4)

_____ I can describe several examples of locomotion. (Ch. 4)

_____ I can describe several examples of growth patterns in living things, including how human head, trunk, and limb proportions change from birth to adulthood. (Ch. 5)

_____ I can give examples of stimulus and response. (Ch. 6)

_____ I can describe examples of migration. (Ch. 6)

_____ I can explain the difference between warmblooded and coldblooded animals and give examples of each. (Ch. 6)

_____ I can identify and describe how some animals and plants are adapted to survive in their environment. (Ch. 6)

_____ I can identify the different parts of a microscope and their functions, and I can make a wet-mount slide to examine cells and their structures. (Ch. 7)

I have also learned to _____

_____

_____

I would like to know more about _____

_____

_____

Signature: _____

Name _____ Date _____ Class _____

Unit 2 SourceBook Activity Worksheet

A Taxing Question: How Can We Classify Living Things?

Complete this activity after reading pages S24–S29 of the SourceBook.

How would you classify these animals? The science of *taxonomy* classifies living things according to their natural relationships. Taxonomists are scientists who practice taxonomy. One method used by taxonomists to classify living things is a branching diagram called a *cladogram*. A living organism is assigned to a certain branch of a cladogram because it has a unique characteristic, or an adaptation. Thus, a cladogram shows how the long-term adaptations that organisms develop in their efforts to survive result in a diversity of living things.

Imagine that you are a taxonomist. Classify the animals illustrated above according to their unique characteristics. To accomplish this task, do the following:

- Complete items 1–3 using the chart below and the cladogram on the next page.
- Classify the animals as instructed in item 4 on the next page.

Animal	Characteristics and adaptations
cat	hair, milk from mother to feed young, teeth for eating meat, retractable claws, ability to purr
seal	hair, milk from mother to feed young, teeth for eating meat
lizard	scaly skin
lion	hair, milk from mother to feed young, teeth for eating meat, retractable claws
horse	hair, milk from mother to feed young

SCIENCEPLUS • LEVEL GREEN 73

Name _____ Date _____ Class _____

SourceBook Activity Worksheet, continued

1. What characteristics do *all* of the animals except the lizard share? Write these characteristics in the space labeled *1* on the cladogram. This point marks the point where all of the other animals split from their common ancestor, the lizard. The other animals occupy a different branch on the cladogram because they have adapted to specific environmental conditions.

2. What characteristic is shared by only *three* of the animals? Write this characteristic in the space labeled *2* on the cladogram.

3. What characteristic is shared by only *two* of the animals? Write this characteristic in the space labeled *3* on the cladogram.

4. What characteristic is exhibited by just *one* of the animals? Write this characteristic in the space labeled *4* on the cladogram.

5. Now classify the animals. To do this, write the name of each animal on the correct branch of the cladogram. (An animal must exhibit all of the characteristics to the left of its branch.) The lizard is already shown on branch *A*. Remember that at least one unique characteristic separates the animals on each branch of the cladogram.

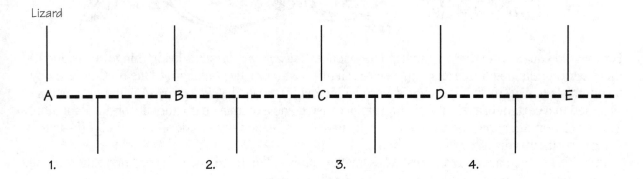

74   UNIT 2 • PATTERNS OF LIVING THINGS

Name _____ Date _____ Class _____

• **Unit 2** SourceBook Review Worksheet

Unit CheckUp, page S63

Concept Mapping

The concept map below illustrates major ideas in this unit. Complete the map by supplying the missing terms. Then extend your map by answering the additional question beneath it.

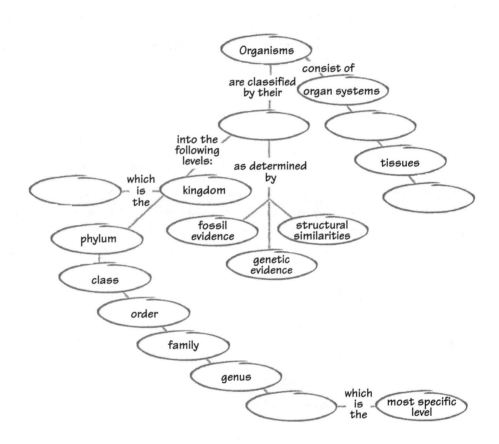

Where and how would you connect the terms *DNA* and *organelles*?

Checking Your Understanding

Select the choice that most completely and correctly answers the following questions.

1. Starfish and humans are members of the same

   a. kingdom. **b.** family. **c.** genus. **d.** species.

2. The material that transmits genetic instructions between generations is

   a. mitochondria. **b.** cytoplasm. **c.** DNA. **d.** testosterone.

SCIENCEPLUS • LEVEL GREEN 75

Name _____ Date _____ Class _____

SourceBook Review Worksheet, continued

3. Which of the following is NOT a function of the human skin?
   a. waste removal
   b. temperature regulation
   c. gas exchange
   d. protection

4. The body system enabling you to read and understand this question is the
   a. optic system.
   b. nervous system.
   c. immune system.
   d. educational system.

5. Chromosomes can be compared with
   a. ladders. b. security fences. c. a shop floor. d. blueprints.

6. Which choice correctly lists the levels of organization from simplest to most complex?
   a. organism, organ system, organ, tissue, cell
   b. cell, tissue, organ, organ system, organism
   c. tissue, organ system, organism, cell, organ
   d. organ, organ system, organism, tissue, cell

Interpreting Illustrations

The accompanying illustration compares the forelimbs of a whale and a human. Identify the corresponding parts, and describe below how these similarities point to an evolutionary relationship between whales and humans.

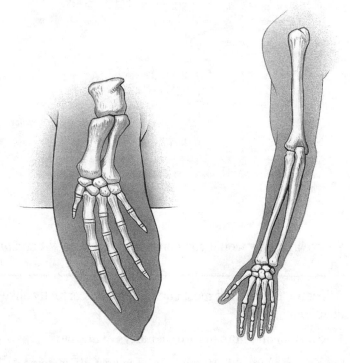

Illustration also on page S64 of your textbook

_____

76 UNIT 2 • PATTERNS OF LIVING THINGS

Name _____ Date _____ Class _____

SourceBook Review Worksheet, continued

Critical Thinking

Carefully consider the following questions, and write a response that indicates your understanding of science.

1. On a late-night television show, a self-proclaimed psychic predicts that four entirely new biological kingdoms will be discovered in the coming year. Is this a realistic prediction? Explain.

   _____
   _____
   _____
   _____

2. In a certain population of foxes, 95 percent are reddish in color, and the rest are solid white. Foxes feed by hunting small animals. Suppose that the climate (which is now fairly moderate with little winter snow) changes and becomes cold and snowy for much of the year. How might this affect the ratio mentioned in the first sentence of this question? Explain.

   _____
   _____
   _____
   _____

3. Certain microorganisms carry out photosynthesis but are also able to take in food from their surroundings and to move about on their own. Speculate how the discovery of organisms such as these helped to do away with the two-kingdom system of classification.

   _____
   _____
   _____
   _____

SCIENCEPLUS • LEVEL GREEN

Name _____ Date _____ Class _____

SourceBook Review Worksheet, continued

4. How does the cell wall determine some of the major defining characteristics of plants? Why would a cell wall be a disadvantage in animal cells?

_____
_____
_____
_____
_____

Portfolio Idea In your ScienceLog, write a story in which you are an organ system, and describe a typical day in your life. Describe what you do, why you do it, how you work with other organ systems, typical problems that might affect you, and so on. Be creative but factually accurate.

Name _____ Date _____ Class _____

Unit 2 SourceBook Assessment

1. The jellylike fluid found within a living cell is composed mostly of
   a. carbohydrates. **b.** proteins. **c.** fats. **d.** water. **e.** oxygen.

2. The members of a _____ have the most characteristics in common.
   a. phylum **b.** kingdom **c.** species **d.** genus

3. According to Darwin's theory of evolution, an archer fish got its ability to squirt water because it
   a. practiced it. **b.** acquired it. **c.** inherited it.

4. Which of the following is NOT one of the three parts of the cell theory?
   a. Only living cells can produce new living cells.
   b. The cell is the basic unit of all living things.
   c. Living cells come from nonliving substances.

5. What level of organization is represented by each of the following?
   a. tree _____
   b. heart _____
   c. blood _____
   d. skeleton _____

6. Identify the cell structure whose function would be most like that of the following parts of a school:
   a. the office _____
   b. the outer walls _____

7. An animal that acquires a specific trait during its lifetime may pass that trait on to its offspring.
   a. true **b.** false

8. Which two organisms would be more closely related: *Ursus arctos* and *Ursus maritimus,* or *Coriandrum sativum* and *Lepidium sativum?*

   _____

9. Discuss the main advantage that specialized cells have over unspecialized cells.

   _____
   _____

SCIENCEPLUS • LEVEL GREEN 79

Name _____ Date _____ Class _____

SourceBook Assessment, continued

10. Can some organs perform more than one function? If so, give an example of an organ with more than one function, and list its functions.

    _____
    _____
    _____
    _____

11. Correct this statement: The proportion of genes you receive from each of your parents depends on which parent's genes were dominant.

    _____
    _____
    _____
    _____

12. What is the function of chlorophyll in plant cells?

    _____
    _____
    _____
    _____

13. List three things that scientists might study to classify a newly discovered life-form.

    _____
    _____
    _____
    _____

14. Classify the following items into two groups: paper clip, apple, pencil, paper, staples, and granola bar. Then explain the characteristics you used to classify the items.

    _____
    _____
    _____
    _____

SourceBook Assessment, continued

15. Why is the scientific system of naming living things necessary? What would biology be like without it?

_____
_____
_____
_____
_____
_____
_____

16. Would it be possible for someone to believe in both the theory of spontaneous generation and the cell theory? Explain your answer.

_____
_____
_____
_____
_____
_____

17. Why would you expect to find a lot of mitochondria in a muscle cell?

_____
_____
_____
_____
_____

18. Match the kingdom on the left with the correct characteristics on the right.

  a. Animalia _____ All make food and have cell walls.

  b. Fungi _____ All are multicellular and ingest food.

  c. Monera _____ Cells do not have nuclei.

  d. Plantae _____ All have nuclei and must absorb food from surroundings.

  e. Protista _____ None have organization above tissue level.

Name _____ Date _____ Class _____

SourceBook Assessment, continued

19. Match the cell structures on the left with the correct function on the right.

   a. chromosome _____ contains instructions for cell processes

   b. cytoplasm   _____ contains DNA and RNA and controls the cell

   c. membrane   _____ limits access to cell and aids in gathering food

   d. mitochondria _____ place where instructions are carried out

   e. nucleus   _____ protein-manufacturing site

   f. ribosome   _____ stores food and waste products

   g. vacuole   _____ turns sugar into energy by cellular respiration

20. Match each element with its corresponding organ system.

   a. sensory neurons _____ circulatory system

   b. bicep   _____ skeletal system

   c. aorta   _____ integumentary system

   d. marrow   _____ muscular system

   e. kidneys   _____ nervous system

   f. epidermis   _____ digestive system

   g. alveoli   _____ excretory system

   h. gallbladder   _____ respiratory system

82   UNIT 2 • PATTERNS OF LIVING THINGS

Name _____ Date _____ Class _____

SourceBook Assessment, continued

21. Complete the following chart to show the similarities and differences between the cellular structure of plants and animals.

Plant cell	Animal cell
cell membrane	
cell wall	
nucleus	
chromosomes	
chloroplasts	
large vacuoles	
ribosomes	
DNA	
mitochondria	

22. Describe what would happen if an earthquake suddenly created a giant gorge, dividing a population of wolves.

_____

_____

_____

_____

_____

23. Give two examples, one from a plant and one from an animal, of a tissue, an organ, and a system.

_____

_____

_____

_____

_____

SCIENCEPLUS • LEVEL GREEN 83

Name _____ Date _____ Class _____

SourceBook Assessment, continued

24. Describe the different types of muscles that make up the muscular system.

_____
_____
_____
_____
_____
_____
_____
_____
_____
_____

25. Explain how human bodies fight diseases internally, and explain why people usually suffer from chickenpox only once.

_____
_____
_____
_____
_____
_____
_____
_____
_____
_____

Unidad 2
Contacto en la casa

Estimado padre/madre de familia,

En las próximas semanas, su hijo o hija va a investigar activamente algunas de las características de los seres vivos, como la capacidad de moverse, de crecer y de responder a estímulos. Cuando los estudiantes hayan terminado la Unidad 2, deberán poder dar respuesta a las siguientes preguntas, para captar las "grandes ideas" de la unidad.

1. ¿Cuáles son algunos de los signos de la vida? (Cap. 4)
2. ¿En qué se distingue el movimiento de las plantas del movimiento de los animales? (Cap. 4)
3. ¿En qué se parecen los modelos de crecimiento de las plantas y los de las personas? (Cap. 5)
4. ¿En qué se diferencian? (Cap. 5)
5. ¿Cuáles son algunas de las diferentes formas de crecimiento? (Cap. 5)
6. ¿Cuál es la importancia de cada una? (Cap. 5)
7. ¿Qué significa la "respuesta a estímulos"? (Cap. 6)
8. ¿Por qué tiene valor esta respuesta? (Cap. 6)
9. ¿Qué pruebas hay de que los animales tienen sentido del tiempo? (Cap. 6)
10. ¿Cuáles son algunas de las adaptaciones de las plantas y los animales a lo largo del tiempo? (Cap. 6)
11. ¿Cómo se adaptan las especies de plantas y animales a lo largo del tiempo? (Cap. 6)
12. ¿En qué se parecen las células de las plantas y las de los animales? ¿En qué se diferencian? (Cap. 7)
13. Describe el proceso para preparar una bandeja con muestras para microscopio. (Cap. 7)

A continuación se mencionan actividades que, si usted quiere, puede practicar con su hijo o hija en la casa.

- Haga que su hijo o hija encuentre en la casa seres vivos y seres inanimados. Pregúntele qué semejanzas y diferencias nota en estas dos clases de cosas. ¿Puede encontrar cosas que ahora no son seres vivos, pero que en otro momento sí lo fueron?
- Hable con su hijo o hija sobre su propio modelo de crecimiento. Trate de contarle sobre los períodos de crecimiento que tuvo él o ella. Use fotos familiares y otros medios audiovisuales para ilustrar estos cambios. Compare los modelos de crecimiento de su hijo o hija con lo que usted sabe sobre su propio crecimiento, cuando usted era niño o niña.

Atentamente,

Unidad 2
Contacto en la casa

Los materiales que aparecen abajo van a ser usados en clase para varias exploraciones de ciencia de la Unidad 2. Su contribución de materiales va a ser muy apreciada. He marcado algunos de los materiales en la lista. Si usted los tiene y quiere donarlos, por favor mándelos a la escuela con su hijo o hija para el

_____.

- ○ algodón
- ○ tazas (pequeñas, de plástico)
- ○ tela o cartón de color oscuro
- ○ cuentagotas (goteros)
- ○ linternas
- ○ pinzas
- ○ botellas de vidrio con tapa
- ○ frascos de vidrio (con tapas; pequeños)
- ○ papel cuadriculado

- ○ cinta para medir (métrica)
- ○ periódicos
- ○ manoplas para el horno
- ○ toallas de papel
- ○ jarras
- ○ envoltorio plástico
- ○ elásticos (grandes)
- ○ arena
- ○ aserrín

- ○ semillas (de crecimiento rápido, como frijoles, alfalfa, rábanos, zanahoria o mostaza)
- ○ fuentes o planchas de horno poco profundas
- ○ cajas de zapatos
- ○ tierra
- ○ alfileres
- ○ pañuelos desechables

Desde ya, le agradecemos su ayuda.

Unidad 2
Las grandes ideas

En la Unidad 2, Patrones de los seres vivos, vas a explorar algunas de las características de los seres vivos, como la capacidad de moverse, de crecer y de responder a estímulos. Estudiarás también la forma en que diferentes animales se han adaptado a su medio ambiente. Al final de la unidad, tendrás la oportunidad de usar un microscopio para examinar muestras de células de plantas y animales. Al leer la unidad, trata de responder a las siguientes preguntas. Estas son las "grandes ideas" de la unidad. Cuando puedas contestar estas preguntas, habrás logrado entender bien los principales conceptos de esta unidad.

1. ¿Cuáles son algunos de los signos de la vida? (Cap. 4)
2. ¿En qué se distingue el movimiento de las plantas del movimiento de los animales? (Cap. 4)
3. ¿En qué se parecen los modelos de crecimiento de las plantas y los de las personas? (Cap. 5)
4. ¿En qué se diferencian? (Cap. 5)
5. ¿Cuáles son algunas de las diferentes formas de crecimiento? (Cap. 5)
6. ¿Cúal es la importancia de cada una? (Cap. 5)
7. ¿Qué significa la "respuesta a estímulos"? (Cap. 6)
8. ¿Por qué tiene valor esta respuesta? (Cap. 6)
9. ¿Qué pruebas hay de que los animales tienen sentido del tiempo? (Cap. 6)
10. ¿Cuáles son algunas de las adaptaciones de las plantas y los animales a lo largo del tiempo? (Cap. 6)
11. ¿Cómo se adaptan las especies de plantas y animales a lo largo del tiempo? (Cap. 6)
12. ¿En qué se parecen las células de las plantas y las de los animales? ¿En qué se diferencian? (Cap. 7)
13. Describe el proceso para preparar una bandeja con muestras para microscopio. (Cap. 7)

Unidad 2
Patrones de los seres vivos

Vocabulario

Adaptation (106)	**Adaptación** característica heredada que aparece con el tiempo y que permite que un organismo sobreviva mejor dentro de un medio ambiente dado
Biorhythm (99)	**Biorritmo** cualquiera de varios ciclos que relacionan hábitos de los organismos como sueño, alimentación, reproducción, migración u otros a los cambios en la luz del sol, la temperatura, las mareas, y otros procesos naturales
Cancer (87)	**Cáncer** un crecimiento perjudícial y descontrolado de los tejidos animales
Cell (113, S36)	**Célula** la unidad más pequeña de los seres vivos
Cell membrane (117, S37)	**Membrana celular** la capa exterior de una célula animal; la parte de la célula que determina qué entra y qué sale de la célula
Cell wall (116, S41)	**Pared celular** la capa exterior de una célula vegetal
Chloroplast (117, S42)	**Cloroplasto** una pequeña estructura que se encuentra dentro del citoplasma de las células de las plantas y de las algas, donde se usa la luz del sol para fabricar azúcar
Coldblooded (104)	**De sangre fría** que tiene temperaturas corporales que cambian de acuerdo a la temperatura de los alrededores
Cytoplasm (117, S37)	**Citoplasma** sustancia parecida a la gelatina que rodea el núcleo de la célula, donde se desarrollan las actividades celulares
Dead (63)	**Muerto** que estaba vivo, pero que ya no presenta ninguno de los signos vitales
Dormancy (104)	**Estado latente** un estado parecido al sueño que puede ser causado por cambios en el medio ambiente
Ectotherm (124)	**Ectotermo** animal que tiene una temperatura corporal muy similar a la de sus alrededores
Ectothermic (104)	**Ectotérmico** que tiene una temperatura que cambia para igualar la temperatura de sus alrededores; también llamado de sangre fría
Endotherm (124)	**Endotermo** animal que mantiene una temperatura corporal bastante constante, independiente de la temperatura de sus alrededores
Endothermic (101)	**Endotérmico** que tiene una temperatura relativamente constante, independiente de la temperatura de sus alrededores
Evolution (125, S30)	**Evolución** proceso por el cual los organismos vivos cambian con el paso del tiempo
Gall (87)	**Agalla** crecimiento dañino en el tejido de las plantas

Vocabulario, continuad

Germinate (81)	**Germinar** brotar; comenzar a crecer de una semilla
Irritability (92)	**Irritabilidad** la capacidad de un organismo de responder a estímulos
Living (63)	**Vivo** que en el momento tiene vida
Locomotion (66)	**Locomoción** la capacidad de un animal de moverse de un lado a otro por sus propios medios
Migration (100)	**Migración** movimiento de animales de un lugar o clima a otro como respuesta a un cambio de estaciones
Mitochondria (117, S40)	**Mitocondrias** pequeñas estructuras que se encuentran en el citoplasma de células vegetales y animales y que sirven como centro de producción de energía para las células
Nonliving (63)	**Inanimado** que nunca ha estado vivo
Nucleus (plural, *nuclei*) (117, S37, S110)	**Núcleo** centro de control de las actividades de vida de una célula; en un átomo, la parte central, que contiene la mayoría de la masa del átomo
Ovary (88)	**Ovario** glándula reproductora femenina que forma huevos y que, en los vertebrados, produce hormonas sexuales
Perspire (102)	**Transpirar** sudar, secretar agua por las pequeñas glándulas en la piel
Regeneration (84)	**Regeneración** crecimiento de nuevas partes del cuerpo para remplazar las que se perdieron o que se dañaron
Reproduction (87)	**Reproducción** proceso por el cual los organismos producen crías de su misma especie; es necesaria para la existencia continuada de una especie
Response (92)	**Respuesta** reacción a un estímulo
Species (87, S25)	**Especie** grupo de organismos que pueden reproducirse entre sí y que se parecen en apariencia, comportamiento y estructura interna; el nivel más específico del sistema de clasificación
Sprout (81)	**Brotar** comenzar a crecer de una semilla; germinar
Stimulus (92)	**Estímulo** cualquier cosa que causa que un ser vivo reaccione
Thermal pollution (127)	**Contaminación termal** tirar aguas servidas calientes en un océano, río o arroyo, lo que daña el medio ambiente por el aumento en la temperatura del agua
Vacuole (117, S40)	**Vacuola** pequeña cavidad dentro del citoplasma de una célula; está encerrada por una membrana y sirve como área de almacenaje, generalmente para alimento, agua o aire
Warmblooded (101)	**De sangre caliente** que tiene una temperatura que permanece relativamente constante, independientemente de la temperatura de los alrededores
Wet mount (115)	**Muestra** muestra delgada de una célula que se coloca en agua entre una bandeja de microscopio y una cubierta

Chapter 4

Name _____ Date _____ Class _____

Exploration 1 Worksheet, continued

Questions

1. What do you think made the animals move when the cover was first removed?

   The animals probably moved in response to the sudden increase in light.

2. Why do you think the animals continued to move?

   Answers will depend on the animals used. The animals probably continued to move in order to get out of or into the light.

3. Which body parts did each animal use for locomotion?

   Answers will depend on the animals used. (For clarity, have students draw and label the different body parts used by the animal.)

4. If all of the animals you observed were in a 2-minute race, which one would win?

   Answers will vary depending on the animals used.

5. What characteristic of this animal would make it the winner of such a race? Students should point out locomotive characteristics that enhance speed. In general, animals that can fly or leap move the fastest.

6. According to your data, which animal would come in last? Can you suggest reasons why?

   Earthworms, snails, and other animals without legs probably move the slowest.

Photos also on page 68 of your textbook

SCIENCEPLUS • LEVEL GREEN   5

Photo/Art Credits

Abbreviated as follows: (t) top; (b) bottom; (l) left; (r) right; (c) center; (bkgd) background.

Photo Credits
Front Cover: (bkgd), Page Overtures. Back Cover: (tl), Tony Stone Images; (bl), Jeff Smith/FotoSmith/Reptile Solutions of Tucson; (bkgd), Page Overtures. Title Page: i (bkgd), Page Overtures; (bl), Jeff Smith/FotoSmith/Reptile Solutions of Tucson. Page 5, HRW photos by John Langford; 13, HRW photo by John Langford; 25, HRW photo by Sam Dudgeon; 44, HRW photo by Sam Dudgeon; 54, Bruce Iverson; 61, Runk/Schoenberger/Grant Heilman, Inc.

Art Credits
All work, unless otherwise noted, contributed by Holt, Rinehart and Winston
Page 7, Morgan Cain & Associates; 9, Lori Anzalone/Jeff Lavaty Artist Agent; 10, Patrick Gnan/Deborah Wolfe Limited Artists' Representative; 11, Walter Stuart/Richard W. Salzman Artist Representative; 16, The Mazer Corporation; 21, Doug Walston; 39, Gary Locke/Suzanne Craig Represents; 40, Guy Wollek; 46, Morgan Cain & Associates; 49, Morgan Cain & Associates; 50, Uhl Studio; 58, Morgan Cain & Associates; 65, Morgan Cain & Associates; 66, Morgan Cain & Associates; 67, Morgan Cain & Associates; 73, The Mazer Corporation; 76, Walter Stuart/Richard W. Salzman Artist Representative; 91, Lori Anzalone/Jeff Lavaty Artist Agent; 92, (t) Walter Stuart/Richard W. Salzman Artist Representative, (b) Patrick Gnan/Deborah Wolfe Limited Artists' Representative; 95, The Mazer Corporation; 103, Gary Locke/Suzanne Craig Represents; 104, Guy Wollek; 107, Uhl Studio; 111, Morgan Cain & Associates; 114, Morgan Cain & Associates; 115, Morgan Cain & Associates; 118, Walter Stuart/Richard W. Salzman Artist Representative.

Answer Keys

Unit 2: Patterns of Living Things

Contents

Chapter 4	90
Chapter 5	96
Chapter 6	102
Chapter 7	106
Unit 2	111
SourceBook Unit 2	117

Name _____ Date _____ Class _____

Exploration 2

Chapter 4
Exploration Worksheet

Solve the Insect Movement Mystery, page 69

Your goal	to learn more about the process of locomotion by determining how an insect moves

You Will Need
- a crawling insect
- a piece of paper

Safety Alert!

What to Do

1. Put a crawling insect on a piece of paper.
2. Watch carefully to see how the insect moves. Decide which of the following things happens:

   a. Both of the front legs move forward at the same time.
   b. Both of the middle legs move forward at the same time.
   c. Both of the back legs move forward at the same time.
   d. All three legs on one side move forward at the same time.
   e. On one side, the front and middle legs move forward at the same time.
   f. On one side, the front and back legs move forward at the same time.
   g. All of the legs move at different times.
   h. Three legs move forward at the same time: the front and middle legs on one side and the back leg on the other side.
   i. Three legs move forward at the same time: the front and back legs on one side and the middle leg on the other side.

3. Record your choice and why you think it is correct.

   Answers could vary depending on the speed of the insect, but (i) is the most accurate response. While an insect is walking, it always has at least three legs supporting its body. The supporting legs are positioned like a triangle. For example, during a step, an insect may move the middle leg on one side with the front and rear legs on the other side. During the next step, the other three legs are moved together.

Name _____ Date _____ Class _____

Exploration 3

Chapter 4
Exploration Worksheet

How's Your Horse Sense? page 69

Your goal	to learn more about the process of locomotion by determining how a horse moves

Below (and on page 69 of your textbook) is a sequence of pictures showing a horse walking. The pictures are arranged to show the order in which steps are taken. Study the set of pictures carefully. Then answer the following questions.

Illustration also on page 69 of your textbook

Questions

1. As the horse walks, how many hooves touch the ground at a time? Are all of the horse's hooves off the ground at any one time?

   At least two hooves usually touch the ground during a walk. At no time are all four hooves off the ground while the horse is walking.

2. Describe the way the horse moves its legs throughout the walking sequence—for example, "right foreleg first, left hind leg second," and so on.

   The walking sequence is right hind leg, right foreleg, left hind leg, left foreleg.

3. Have you ever seen a baby crawl? How does the horse's walking gait compare to the movements of a crawling baby?

   They are similar in that both alternate limbs from front to back and from side to side.

Exploration 3 Worksheet, continued

Comparing Structures

1. The illustrations below show a horse's forelimb and its human counterpart. Label the equivalents of the fingernails, fingers, hand, wrist, forearm, elbow, and upper arm on the diagram of the horse's leg.

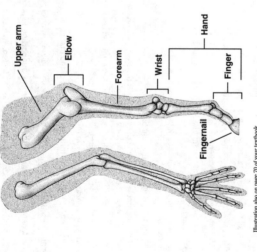

Illustration also on page 70 of your textbook

2. How is each structure in each forelimb well suited to its function?

   The structure of the human wrist and hand is very complex. Four fingers and a thumb are used to grasp objects, while the wrist joint allows for a great deal of flexibility. The thick bones of the horse's upper leg support a lot of weight, while the hoof is flat and provides traction.

   _____
   _____
   _____
   _____
   _____
   _____

EXPLORATION 4
Chapter 4
Exploration Worksheet

Listening to an Earthworm! page 72

| Your goal | to determine how an earthworm moves |

You Will Need
- an earthworm
- a piece of stiff paper
- 10 mL of water

What to Do

1. Put an earthworm on a piece of stiff paper. Can you hear it make any noise as it moves?

2. Hold the paper up level with your eyes, and try to look between the animal and the paper. Do you see the bristles? Rub your finger back and forth along the lower side of the earthworm. Do you feel the bristles? When the stiff bristles are extended, they hold the animal in place on the ground. When the bristles are pulled in, the animal can slide along. When the earthworm moves along the paper, its bristles sometimes make a scraping noise.

3. Notice the rings on the body of the worm. The inside of an earthworm's body is divided into sections called *segments*, which show up on the outside as rings. There are four pairs of bristles for every segment. How many bristles does your earthworm have?

   Answers will vary depending on the segments used.

4. Now put the earthworm on damp paper. Watch it move. How does the earthworm use its bristles to propel itself?

   Students may note that the bristles are used to gain traction on the surface below the earthworm.

   Like snakes, earthworms have muscles. How does the earthworm use its muscles to move?

   The earthworm uses its muscles together with its bristles to move the segments of its body.

5. Look at the illustration. Locate the muscles. These muscles are very strong. One set of muscles goes around the worm in rings.

   An inside and lengthwise look at an earthworm

   Lengthwise muscles (these go the length of the worm)
   Skin
   Circular muscles (these go around the body)
   Bristles

   Illustration also on page 72 of your textbook

Name _____ Date _____ Class _____

Chapter 4
Review Worksheet

Challenge Your Thinking, page 73

1. Wanted: Nonliving or Living

What characteristics do all living things have in common?

All living things share characteristics such as growth, reproduction, food consumption, movement, response to stimuli, ingestion, respiration, and excretion.

Sometimes a nonliving thing has a characteristic of a living thing. How many examples of this can you list? Make a list. Can you top 20 examples?

Examples of nonliving things that exhibit certain signs of life include machines that move and mechanical devices that respond to stimuli, such as smoke detectors or burglar alarms.

If a nonliving thing has a characteristic of something that is alive, why is it classified as nonliving?

For something to be alive, it must have *all* of the characteristics of living things.

2. Classify It

A biologist made the following classification of all objects:

a. living—having all of the signs of life
b. dead—having once had all of the signs of life
c. nonliving—having never had all of the signs of life

Where does a wooden chair fit in? How about other things—water, mushrooms, a pie? Classify each of the following into one of the three groups: oyster, moss, yeast cake, salt, sugar, bones, hibernating bear, pencil, wool sweater, seaweed, sponge, pearl, volcano, kernel of corn, clams, barnacle on a rock, bean seed, baked beans, freshly picked strawberries, pine cone, lichens on a rock, electric fan, cactus, paper.

Record your answers in the chart on the next page.

Name _____ Date _____ Class _____

Exploration 4 Worksheet, continued

6. Now locate the muscles that run lengthwise along the worm's body. The two sets of muscles work against each other.

a. What happens when one set of these muscles contracts and the other set relaxes?

When one set of muscles contracts and the other set relaxes, the earthworm moves forward.

b. When the ring muscles contract, what happens to the length of the segment?

Each segment lengthens when the ring muscles contract.

c. When the lengthwise muscles contract, what happens to the length of the segment?

Each segment shortens when the lengthwise muscles contract.

7. See if you can piece together this information about the muscles and bristles to write a description of how earthworms move.

Sample description: When one set of muscles contracts and the other set relaxes, the earthworm moves forward. Each segment lengthens when the ring muscles contract and shortens when the lengthwise muscles contract. The bristles give the earthworm traction.

8. Compare your description with Dan's description on page 72 of your textbook.

Comparisons will vary but should be clear and complete. Students will most likely comment on the thoroughness of Dan's description compared to their own.

Name _____ Date _____ Class _____

Chapter 4 Review Worksheet, continued

Living	Dead	Nonliving
Mushrooms, oyster, moss, yeast cake, hibernating bear, seaweed, sponge (if natural and still living in water), kernel of corn (contains an embryo), clams, barnacle, bean seed, seeds in the strawberries, seeds in the pine cone, lichens, cactus	Pie, wooden chair, bones, the wood in the pencil, wool sweater, baked beans, paper	Water, salt, sugar, the lead in the pencil, pearl, volcano, electric fan

3. Curious Questions

Your class has just been asked to write the section called "Curious Questions" in the book *Mr. Know-It-All's Science Facts*. Choose one question from the list below. The question should be answered clearly in a paragraph of 150 words or less and should be understandable to an 11- or 12-year-old. If an illustration would help, draw one! Continue in your ScienceLog if necessary.

a. How do you tell a plant from an animal?

b. In how many different ways can animals move?

c. How does an insect use its six legs to move?

d. How does a horse use its four legs to move?

(a) Unlike animals, most plants make their own food, and most contain pigments for making food. Also, plants are typically unable to move from place to place; they can move only parts of their body.

(b) Swimming, burrowing, walking, running, hopping, crawling, soaring, and flying are some examples of animal locomotion.

(c) Insects usually walk by moving the middle leg on one side at the same time that they move the front and hind legs on the other side.

(d) A horse walks by alternating its legs from front to back and from side to side. One possible sequence could be the following: left rear, left front, right rear, right front.

---

14 UNIT 2 • PATTERNS OF LIVING THINGS

Name _____ Date _____ Class _____

Chapter 4 Review Worksheet, continued

4. That's Life?

You are a scientific explorer, and you have just come across the mysterious blob shown on page 74 of your textbook. Do you think it is alive? What characteristics would you look for? Write down a series of steps you would take to answer these questions.

Students should look for a combination of the following characteristics: growth, reproduction, food production or consumption, movement, response to stimuli, having a life cycle, ingestion, digestion, respiration, and excretion. The steps that the students suggest should allow for careful observation of the subject and should include exposing the subject to a variety of stimuli. For more information about the photograph, see page 74 of the Annotated Teacher's Edition.

5. It's All Connected

Create a concept map to show how the following words and phrases are related: *living things, movement, human, scales, bristles, limbs, snake, animal,* and *earthworm*. To create the map, arrange these items in a logical way, and use lines and connecting phrases to link them.

Accept all reasonable responses. For instance, between the words *animal* and *living things*, students might put "is an example of." Between *limbs* and *human*, students might put "are used for locomotion by a."

SCIENCEPLUS • LEVEL GREEN   15

Name _____ Date _____ Class _____

Chapter 4 Assessment

Short Responses

1. Do you agree or disagree with the following statement? Explain your reasoning.

   Plants cannot move because they do not have the necessary limbs for locomotion.

   Students should disagree. Although plants cannot move from place to place, they do move in response to changes in their environment.

2. Explain why food production is not one of the universal signs of life.

   Sample answer: Some living things produce their own food, like plants, but others do not, like animals. Unlike some of the other signs of life (food consumption, growth, reproduction, etc.), food production isn't a characteristic of *all* living things.

Illustration for Interpretation

3. Use the illustration below to answer the questions on the next page.

Name _____ Date _____ Class _____

Chapter 4 Assessment, continued

a. List the living, nonliving, and dead things that you see in the illustration.

Living	Trees, grass, human, dog, people in the airplane
Nonliving	Air, mountains, stereo, stereo batteries, fork, plate, airplane, any portion of the man's clothing that is synthetic, the nails that hold the picnic table together
Dead	The wood that makes up the picnic table; any portion of the man's clothing that is made from natural fibers such as cotton, silk, or leather; the organic materials from which the airplane fuel is composed

b. What characteristics of living things help you distinguish them from nonliving things?

   Sample answer: Living things grow, reproduce, and respond to stimuli; nonliving things do not.

Short Essay — CHALLENGE

4. Earthworms and snakes have similar means of locomotion. Compare and contrast each animal's method of getting from place to place, describing the techniques of each one and pointing out their similarities and differences.

   Sample answer: An earthworm has bristles on the underside of its body that can be extended to hold it in place or retracted to allow motion. The scales of snakes can also be extended or retracted. However, a snake actually uses its scales to move. The scales catch rough surfaces, and the snake's muscles (which are attached to the scales) move it forward. When an earthworm's bristles are retracted, its two sets of muscles alternately contract and lengthen to move it forward. The reason that a snake has difficulty moving on slick surfaces is similar to the reason that an earthworm has bristles. An earthworm's body is very smooth, and it needs its bristles to hold it in place on the ground when not in motion.

Name _____ Date _____ Class _____

Chapter 5
Activity Worksheet

Growth Trivia

This activity corresponds to Lesson 2, which begins on page 80 of your textbook.

Who can find the most answers? Give yourself 48 hours to answer the following questions:

1. How long does it take a puppy to become full grown?	18 months to 3 years
2. Does the life span of an animal have anything to do with how long it takes it to grow up?	Yes, the longer it takes an animal to grow up, the longer its life span is.
3. A dog at 14 years of age is physically as old as a human at 98 years. How many years of a human's life equals 1 year of a dog's life?	7 years
4. How long does a baby elephant remain dependent on its mother?	4 years
5. At what age is an elephant full grown?	About 20 years
6. On average, how long does an elephant live?	About 100 years
7. Name two mammals that grow hair only after they are born.	Some examples are kangaroos, opossums, mice, and rats.
8. List two animals that, when they are born, are basically on their own.	Some examples are snakes, frogs, toads, turtles, lizards, insects, and fish.
9. How does a hen show signs of old age?	One example is a decrease in egg production.
10. How does a lobster or a grasshopper—both of which are covered by hard shells—grow?	By a process called molting
11. What animals change completely as they grow, so that the young do not look like the adults?	Some examples are caterpillar to pupa to butterfly (or moth), nymph to dragonfly, tadpole to frog (or toad), and alevin to fry to salmon.
12. How are chickens different from each other—even those that look alike?	Each has a different number of feathers.
13. How many hairs are on your head?	Approximately 120,000
14. As we grow, we often grow differently; but what is different about every human, even at birth?	Our fingerprints
15. How tall is, or was, the tallest human?	268 cm

SCIENCEPLUS • LEVEL GREEN 19

Name _____ Date _____ Class _____

Chapter 4 Assessment, continued

CHALLENGE 2
Numerical Problem

5. An earthworm has 4 pairs of bristles on each of its segments. Suppose that you find an earthworm with 64 segments.

   a. How many individual bristles does the earthworm have? Show your work.

   64 segments × 4 pairs/1 segment × 2 bristles/1 pair = 512 bristles

   b. You find another earthworm that is three-fourths of the size of the first earthworm. How many individual bristles does this earthworm have? Show your work.

   64 segments × 3/4 = 48 segments

   48 segments × 4 pairs/1 segment × 2 bristles/1 pair = 384 bristles

   c. Why might a scientist be interested in the number of bristles on an earthworm?

   Accept all reasonable responses. Sample answer: Every living thing plays an important role in the environment. For this reason, scientists are interested in finding out as much as possible about every species, including earthworms. Earthworms are especially interesting to scientists who are concerned about soil and agriculture. This is because earthworms move through the soil, mixing and loosening it and promoting drainage. They also help to decompose organic material in the soil, which produces more nutrients for growing plants. By studying the number of bristles on earthworms, scientists are able to classify different types of earthworms and to learn more about the specific role of each type.

18 UNIT 2 • PATTERNS OF LIVING THINGS

Name _____ Date _____ Class _____

EXPLORATION 1
Chapter 5
Exploration Worksheet

Changing Conditions, page 81

Cooperative Learning Activity

Group size	3 to 4 students
Group goal	to test different conditions for growing seeds and to develop an understanding of the term *variable*
Individual responsibility	Each member of your group should choose a role such as materials coordinator, recorder, leader, or investigator.
Individual accountability	Each group member should be able to define the term *variable* and explain why it is important to identify variables when conducting an experiment.

Imagine that you are a botanist working for a seed company. You are responsible for testing different conditions for growing seeds and for developing directions to go on the label.

You Will Need

- 10–20 quick-growing seeds, such as mung bean, alfalfa, radish, carrot, or mustard seeds
- a variety of seed-growing materials provided by your teacher

What to Do

1. Divide your seeds into several small groups. You should have more than one seed in each group.
2. Discover what conditions are needed to produce the best results.
3. In your ScienceLog, make a report for the company files, and write a draft of the instructions for the seed package. In your report, include the following information: the sets of conditions you tested, what happened to each group of seeds, and which conditions were the best.
4. One word you might find useful in writing the report is *germinate*. This word means "to begin to grow from a seed."
5. In performing your tests, be sure to change only one variable. Remember that a variable is a condition that can or does change. Variables that you could change include temperature, moisture, and amount of light. All other variables should remain the same.
6. Some methods for germinating seeds are shown on page 81 of your textbook. Use the same type of seed for each set of conditions. Try to include bean, carrot, or radish seeds. You can read about the method one group of students used on page 81. They used several seeds for each condition. Why is this important?

   It is important to use several seeds for each condition because an individual seed may not germinate due to damage or a genetic flaw.

7. Before you begin, make a prediction in your ScienceLog about what you think the result of each experiment will be.

20 UNIT 2 • PATTERNS OF LIVING THINGS

Name _____ Date _____ Class _____

Graphing Practice Worksheet, continued

Questions

Jane is ready to begin planting. Can you help her? Use the graph to answer the following questions:

1. Will any of the seeds freeze in region 1? in region 2?

   Seeds A and C will freeze in region 1; none of the seeds will freeze in region 2.

2. Will any of the seeds shrivel in region 1? in region 2?

   Seeds A and B will shrivel in region 1; seed B will shrivel in region 2.

3. Jane wants to grow plants all year long in region 1. Which seed should she use?

   None of the seeds can be used to grow plants all year long in region 1.

4. Jane wants to plant seeds in region 2 in March and harvest the plants in July. Which seed should she use? Why?

   Seeds A and C could both be used, but Jane should plant seed A because the average temperature stays closer to seed A's ideal temperature for growth.

5. When do the two regions have the same average temperature? When do the average temperatures differ the most?

   They are the same in May and October; they differ the most in March and July.

6. List at least four factors other than temperature that might influence how well Jane's plants grow.

   Answers will vary. Possible factors include the amount of sunlight the crops receive, the condition of the soil, the amount of moisture received, and the presence or absence of insects or disease.

22 UNIT 2 • PATTERNS OF LIVING THINGS

Name _____ Date _____ Class _____

Chapter 5
Review Worksheet

Challenge Your Thinking, page 89

1. An Eggsercise

Take a look at the photograph of the chicken egg below. Use the photograph and what you have learned in this chapter to answer the following questions:

Photo also on page 80 of your textbook

a. What is the purpose of the eggshell?
The eggshell protects the developing embryo. (It is also porous to allow for gas exchange.)

b. What is the purpose of the yolk? How long does it have to last?
The yolk is the food for the developing embryo. It must last until the chicken hatches.

c. Where in the egg would the young chick have developed if the egg had been fertilized?
The young chick would have developed from the yolk.

d. How does the structure of an egg protect a developing chick?
The shell provides protection against external dangers. The embryo is surrounded by the egg white, which cushions the embryo during development and protects the embryo against disease-causing organisms that penetrate the shell.

SCIENCEPLUS • LEVEL GREEN 25

Name _____ Date _____ Class _____

Chapter 5
Activity Worksheet

Growth Puzzle

Tackle this crossword puzzle after you've finished Lesson 2, which begins on page 80 of your textbook.

Each paragraph on the following page is a summary of material you have studied so far. You will notice, though, that key words are missing. In their place is a number followed by the word *across* or *down*. When you think you have the right word for a blank, see whether your answer fits into the boxes for that number across or down.

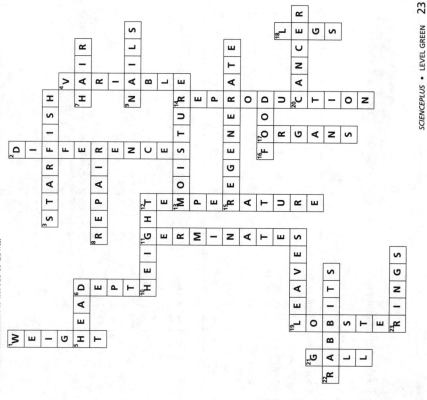

SCIENCEPLUS • LEVEL GREEN 23

98 UNIT 2 • PATTERNS OF LIVING THINGS

Chapter 5 Review Worksheet, continued

c. How can you tell the ages of plants and animals?
   In animals, age can be determined by the size, shape, and proportions of body parts; in plants, age can be determined by the total size of the plant, the leaf size, the circumference of the stems and branches, and the number of rings in the trunk.

d. Can an animal lose a part of its body and grow a new one to replace it?
   Some animals, including planaria, lizards, salamanders, crabs, lobsters, starfish, and tadpoles, can regenerate body parts.

e. Do all parts of humans grow at the same rate?
   Different parts of the body grow at different rates, depending on the stage of growth.

4. **Parenting Skills**

   Animals with backbones are called *vertebrates*. They develop from fertilized eggs. Their ways of reproducing, however, show many differences. In fact, you could sort them into groups by answering a few questions about their fertilized eggs and what happens to them. Look at the chart on page 90 of your textbook.

   a. The letters *A* through *F* in the diagram represent headings that could be in the form of questions. For each letter, supply the question that is answered by that section of the chart. For example, the question for *A* could be, Where does fertilization take place?

   Answers will vary. Sample answers: (B) What type of eggs are produced? (C) For eggs with shells, are the shells brittle or soft? (D) Does the fertilized egg develop inside or outside the female? (E) Are the eggs cared for? (F) For how long and in what way are the young cared for?

Chapter 5 Review Worksheet, continued

2. **Check It Out**

   Look at the bar graph illustrating the height of students in Ada's class, and then answer the following questions. Ada is 130 cm tall.

   a. How many students are in Ada's class? **24**

   b. How many boys are shorter than Ada? **2**

   c. How many girls are as tall as Ada or taller? **10 (11 including Ada)**

   Height of students (cm) — bar graph with Boys and Girls, bins 120–129, 130–139, 140–149, 150–159; Number of students on y-axis (0–10).

3. **Inquiring Minds**

   Add the answers to the following "Curious Questions" that you worked on earlier in the unit.

   a. Why does a baby have such a big head?
      The brain develops rapidly because of its immediate importance in carrying out bodily functions. A baby, therefore, has a large head with respect to total body size.

   b. Is bigger always better?
      Answers will depend on students' opinions. In general, when an organism is bigger, it usually requires a greater amount of energy to sustain itself.

Chapter 5
Assessment

Word Usage

1. Show that you understand the following terms by using them in one or more sentences:

   a. regeneration, cancerous growth, starfish, tumor, renewal, continual growth, skin, hair

   Sample answer: *A tumor is an example of a cancerous growth. A starfish demonstrates the process of regeneration by replacing a lost arm. Renewal is a form of growth that repairs damage. For example, a blister shows that the skin is repairing itself. Continual growth is a form of growth that constantly replaces structures like hair that wear away.*

   b. head, proportion, size, adult, birth

   Sample answer: *At birth, a person's head takes up a larger proportion of total body size than it does when the person is an adult.*

Correction/Completion

2. The statements below are either incorrect or incomplete. Make the statements correct and complete.

   a. The cells of a person's skin may start growing uncontrollably after the person has spent 3 hours a day in the sun for the past 15 years. This kind of growth is known as regeneration.

   The cells of a person's skin may start growing uncontrollably after the person has spent 3 hours a day in the sun for the past 15 years. This kind of growth is known as *cancer*.

   b. The part of the body that grows the fastest from birth to the age of 5 is _____.

   the brain

Chapter 5 Review Worksheet, continued

b. Based on the information in the chart, which of the animals named in the far right column do you think would have the best chance to survive and to grow up? Which would be the least likely to survive? Give reasons for your choices.

The human, dog, and eagle probably have the best chances of survival. The salmon, frog, and turtle are probably least likely to survive. In general, animals with internal fertilization and parental care of eggs and young have the best chance to survive and to grow up.

c. Why do you think a frog produces many more eggs at one time than a bird does?

If a frog lays many eggs, there is a better chance of many young surviving to adulthood. A bird, however, is limited by the number of eggs that it can care for.

d. Which do you think produces more eggs at one time, a fish or a turtle? Why?

Fish produce more eggs than turtles because fish eggs are fertilized externally and are not surrounded by a shell.

Name _____ Date _____ Class _____

Chapter 5 Assessment, continued

Short Response

3. How is regeneration different from reproduction?

   Sample answer: Regeneration is the process by which an organism recreates a missing part of its own body. Reproduction, on the other hand, is the process by which an organism or organisms create another organism that exists independently of the parent organism.

CHALLENGE

Graph for Interpretation

4. Use the following graph showing the heights of three generations of the Kapasi family to answer the questions that follow.

   X = first generation
   ☆ = second generation
   ○ = third generation

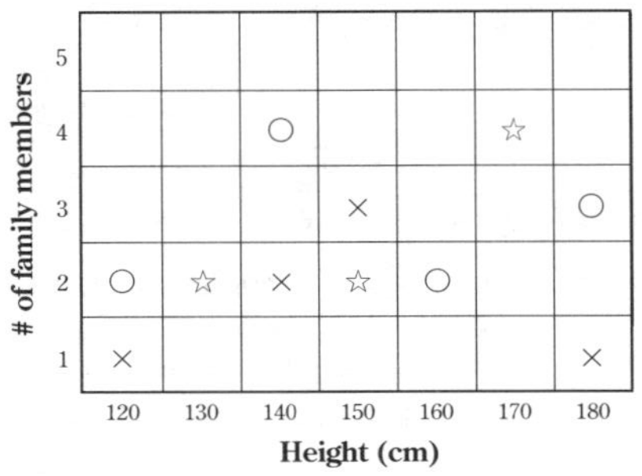

   [Graph: # of family members (y-axis, 1–5) vs Height (cm) (x-axis, 120–180)]

   a. How many members are in each of the first and second generations of the family? In the first and second generations combined?

   The first generation has 7 members, and the second generation has 8 members, so there are a total of 15 members in the first and second generations combined.

Name _____ Date _____ Class _____

Chapter 5 Assessment, continued

   b. What is the average height of each generation? Of all three generations? Show your work.

   First generation:
   [120 cm + (2 × 140 cm) + (3 × 150 cm) + 180 cm] ÷ 7 = 147 cm
   Second generation:
   [(2 × 130 cm) + (2 × 150 cm) + (4 × 170 cm)] ÷ 8 = 155 cm
   Third generation:
   [(2 × 120 cm) + (4 × 140 cm) + (2 × 160 cm) + (3 × 180 cm)] ÷ 11 = 151 cm

   All three generations: (147 cm + 155 cm + 151 cm) ÷ 3 = 151 cm

   c. Can you make any comments about height within a family based on your observations?

   Answers will vary, but students may suggest that if a family's average height remains fairly consistent with different generations, it is possible that height could be passed down from one generation to another.

CHALLENGE 2

Short Essay

5. How might the continual growth of certain parts of an organism be beneficial? Mention at least one example in your answer.

   Sample answer: Continual growth can be beneficial to an organism in terms of protection, continued use, and survival. Because humans use their hands so much, continual fingernail growth protects fingers and hands from any damage that could occur and ensures the continued use of the fingernails if they should occasionally break or tear. Beavers build dams by cutting down trees and gnawing pieces of wood. This process wears down their teeth. However, because their teeth grow continually, beavers are able to continue building and repairing their homes.

Exploration 3

Which Loses Heat Faster—A Mouse or a Mountain Lion? page 103

Your goal to decide if the size of an organism affects the rate of heat loss

Safety Alert!

You Will Need

- 2 glass bottles (with lids), one of which can hold twice as much as the other
- 2 alcohol thermometers
- a pitcher
- hot water from the tap
- oven mitts

What to Do

1. Fill each bottle with hot water at the same time.
2. Record the temperature of the water in each bottle.
3. Measure the temperature of the water in each bottle every 5 minutes for 30 minutes. **Caution: Be careful not to burn yourself with the hot water. Use oven mitts as necessary to handle the bottles of hot water.**

Was there any difference in the rate of cooling of the small and large containers? Try to explain the difference.

Yes. A good explanation for the difference in the rate of heat loss will take into account the fact that although the larger volume of water contains twice as much heat energy as the smaller volume, the surface area of the larger bottle is not twice that of the smaller bottle. Therefore, the rate of cooling in the larger "animal" is slower than that in the smaller one.

Time (min.)	Temp. of small bottle	Temp. of large bottle
5		
10		
15		
20		
25		
30		

Does the result support your prediction about the mouse and the mountain lion?

Student answers will vary according to their predictions but should conclude that the rate of cooling in a larger animal will be slower than that in a smaller animal. In other words, heat content is a function of volume, but heat loss is a function of surface area.

Exploration 1 Worksheet, continued

Questions

1. a. Are all parts of an earthworm equally sensitive to touch?
   No, some parts are more sensitive to touch than others.
   b. If there is a difference, which parts respond more to touch than others?
   The sides and the back end seem to be more sensitive to touch than the front end.

2. a. Which part of the worm seems most sensitive to light?
   The front part of the worm seems most sensitive to light.
   b. Did the worm move toward the light (positive response) or away from the light (negative response)?
   The worm responded negatively to the light.

3. Give two reasons why earthworms are not usually found on the surface of the ground.
   Earthworms are not usually found on the surface of the ground because their food source is found in the soil. Also, earthworms are more vulnerable to predators on the surface.

4. When are you likely to find earthworms on the surface? Why are they there at that time?
   Worms are likely to be found on the surface at night because there is no sunlight. Worms also take in oxygen through their moist skin; after a hard rain, worms will come to the surface because there is not enough oxygen available in the water-soaked soil.

5. How could the earthworm sense the vinegar without touching it?
   The earthworm can sense certain chemicals in the air, a process that is similar to our ability to smell.

6. How are the earthworm's responses useful for its way of life?
   The earthworm's negative response to touch might help it escape from predators. Also, because earthworms respond negatively to light, they remain underground during the day. This keeps them from drying out in the sun and keeps predators from seeing them.

Name _____ Date _____ Class _____

**Chapter 6
Activity Worksheet**

A Warmblooded Puzzle

Complete this activity after finishing Lesson 3, which begins on page 104 of your textbook.

For each item in list I, select the best matching item in list II. (You may have to do a little research on a few of the items.) Enter the letter in the appropriately numbered box in the puzzle square. One letter is used more than once. When you have completely filled the boxes, you will discover a "warmblooded" message!

List I
1. warmblooded winter inhabitant
2. migrates south in winter
3. dormancy
4. panting
5. feather fluffing
6. sweating
7. coldblooded
8. banding
9. rabbit ears

List II
A. robin
E. method of tracking migrating birds
H. slowed breathing and heartbeat
O. way of cooling off
R. way of keeping warm
T. Blood gives off heat through these.
U. Body temperature matches that of surroundings.
Y. black-capped chickadee

1 Y	2 A	3 H
4 O	5 R	6 O
7 U	8 E	9 T

UNIT 2 • PATTERNS OF LIVING THINGS

Name _____ Date _____ Class _____

**Chapter 6
Review Worksheet**

Challenge Your Thinking, page 110

1. An Animal's Tale (Tail?)

Answers will vary. Encourage students to choose animals from various environments around the world.

Each member of the class should choose a specific animal and tell a story, from that animal's point of view, about its preparations for winter. If you choose to be an animal that migrates, tell the class how, when, and where you migrate and the reasons for your migration. If you choose to be an animal that stays put, describe your preparations. Record your story in your ScienceLog. You will have to do research. Look in books on the subject, and ask informed people. Try to make your story about 150 words long.

Illustration also on page 110 of your textbook

2. I Can't Say No to Sweets

In your ScienceLog, describe how you would test the responses of ants to the stimuli of sugar, moisture, and heat. Use labeled diagrams. Make a prediction about what the ants' responses will be.

Answers will vary. Look for logical reasoning in student responses. Students might choose different types or colors of candy and place the candy on an ant mound, for instance. Student observations should describe the activity of the ants and the variables they tested. Remind students to limit their testing to one variable at a time.

SCIENCEPLUS • LEVEL GREEN

Chapter 6 Review Worksheet, continued

3. Good Reflexes

Look at the picture below. Pick out all of the examples of stimuli and responses to these stimuli that you can find. Determine whether each response is positive (+) or negative (−), and record your answers in the following chart:

Sample answers:

Stimulus	Response	+	−
candy in store	child reaches	X	
child reaches	woman pulls child away		X
cat	dog chases cat	X	
dog	cat flees from dog		X
dog and cat in front of car	driver screeches to a halt		X
sunshine	man puts on sunglasses		X
truck backing up	boy on bicycle waits	X	
heat and sunshine	man on the bench wipes brow		X
red light	car stops		X

Illustration also on page 111 of your textbook

Chapter 6 Review Worksheet, continued

4. Strange Behavior

Read the conversation below. Write at least two possible explanations for each animal behavior mentioned.

Pamela: I see ants going up and down the stems of our rosebush. Why do they do that?

Students' explanations should describe the different behaviors as responses to one or more stimuli.

Mike: I have a mystery too. A lot of crows fly into the trees behind our apartment building late in the afternoon each day.

Ani: Some kind of animal dug up big patches of our lawn last night.

Raoul: There's a bird that pecks at one of our windows early every morning. Why does it do that?

Now choose one of the behaviors, and explain how you would test one of your explanations.

The students' explanations should include two separate variables that the students can test. Manipulating one variable at a time will allow students to look for changes in the animals' behavior that might support or refute their explanations.

Chapter 6 Assessment

Word Usage

1. Show that you understand the following terms by using them in one or two sentences:

   a. descendants, distance, humpback whales, migrate, return

   Sample answer: Humpback whales travel a great distance when they migrate. In years to come, their descendants will return to the same locations.

   b. clocks, daylight, schedule, animals

   Sample answer: Some animals maintain a regular schedule, as if they had internal clocks, even when deprived of daylight.

Short Response

2. Explain how negative responses help an organism.

   Sample answer: A negative response of an organism to a certain stimulus is a kind of defense mechanism that protects it from predators or other harmful aspects of its environment. The organism develops a pattern of negative responses over time to ensure its survival as well as to perpetuate the existence of the species.

Correction/Completion

3. The statements below are either incorrect or incomplete. Make the statements correct and complete.

   a. Animals who live in arctic climates have long, thin limbs in order to keep themselves warm.

   Animals who live in arctic climates have small, short limbs that keep more blood in the inner parts of the body in order to keep themselves warm.

Chapter 6 Assessment, continued

   b. Structural adaptations in species develop rapidly so that animals can respond to long-term changes in their environments.

   Structural adaptations develop over long periods of time and continue to develop as conditions change so that animals can adapt to long-term changes in their environments.

4. Imagine a swampy, tropical habitat where annual temperatures range from 25°C to 30°C. The area receives about 250 cm of rainfall each year and is home to many plants and animals, including fish, frogs, snakes, rodents, birds, and alligators.

   Design your own warmblooded, meat-eating animal to live in this environment. On a separate sheet of paper, draw a picture of your animal. Then add labels to indicate what structures the animal has for locomotion, catching prey, protecting itself from predators, and protecting itself from harmful elements in the environment (such as direct sunlight).

CHALLENGE 1
Answering by Illustration

Answer to 4 (above): Answers will vary but should demonstrate an understanding of the relationship between structure and function in animals and should be compatible with the environment described in the question.

5. Consider a shark and a dog. Describe the different ways they respond to their individual environments over the course of a year.

   Sample answer: A shark's body temperature rises with the temperature of its surroundings because it is an ectothermic animal. The body temperature of an endothermic animal, like a dog, remains constant, so the animal must develop a way to remove heat from its body. Dogs breathe rapidly, or pant, to keep cool. The faster a dog pants, the more heat leaves its body. As it becomes colder outside, the shark's body temperature drops with the water temperature and its body functions, like heart and breathing rates, slow down. In the coldest part of winter, a shark will enter a state of dormancy to protect itself from extreme cold. Since a dog's body temperature remains the same during cold weather, it will grow a thicker coat and eat more than usual in order to keep warm.

CHALLENGE 2
Short Essay

Name _____ Date _____ Class _____

Chapter 7 — Exploration Worksheet

EXPLORATION 1

Looking at Cells, page 115

Your goal to prepare a wet mount for a microscope

When you examine the cells of a plant or animal through a microscope, the sample you use must be very thin so that light can shine through it. Also, the sample must be kept moist. The illustrations on page 115 of your textbook show how to make a **wet mount**.

You Will Need

- a microscope
- microscope slide(s)
- coverslip(s)
- an eyedropper
- a small piece of newspaper with a letter *e*
- forceps
- water
- tissue

What to Do

In this activity, you will see that things look different, not just larger, under a microscope.

Making a Wet Mount

1. Put a drop of water on a slide.
2. Place the piece of newspaper with the letter *e* on the water.
3. Lean a coverslip against a pencil. Slowly lower the coverslip, and finally remove it. Avoid getting air bubbles under the coverslip. Absorb excess moisture around the coverslip with tissue.
4. Now view the slide through the microscope, first with the low-power lens and then with the high-power lens. In your ScienceLog, draw what you see.

Analyze Your Work

1. If you move the slide to the left, in which direction do the objects in the field of view move?

   To the right

2. When you change from the low-power lens to the high-power lens, how does your field of view change?

   The field of view becomes smaller due to the higher magnification.

> To calculate total magnification, multiply the magnification of the eyepiece lens by the magnification of the objective lens you are using. A $10\times$ eyepiece lens and a $10\times$ objective lens will give a total magnification of $100\times$.

Watch Out for These!

Dark, round circles are air bubbles.

Dark, jagged lines are coverslip edges.

Thin, irregular lines are caused by water drying up.

Name _____ Date _____ Class _____

Chapter 7 — Resource Worksheet

How to Use a Microscope, page 114

The instrument that allows us to see cells and examine small objects is the microscope. As you read the instructions for its use on page 114 of your textbook, try to match the names of the parts of the microscope (italicized below) with the labels on the diagram. The diagram below shows only one example of a microscope. The microscope that you use in your own classroom may be slightly different. For example, it may have a mirror instead of its own light source.

light source, base, eyepiece, tube, stage, diaphragm, objective lenses, nosepiece, coarse-adjustment knob, fine-adjustment knob

a. **Eyepiece**
b. **Coarse-adjustment knob**
c. **Fine-adjustment knob**
d. **Base**
e. **Tube**
f. **Nosepiece**
g. **Objective lenses**
h. **Stage**
i. **Diaphragm**
j. **Light source**

Photo also on page 114 of your textbook

Name _____ Date _____ Class _____

Exploration 2 Worksheet, continued

5. In your ScienceLog, draw all of the features that you see in one cell. Label the cell and the **cell wall** (the thick casing of the cell).

6. Clean and dry the slides and coverslips.

7. Look at the prepared slide of material taken from the inside of a person's cheek. Notice the color due to the stain used. Instead of iodine, other stains are often used.

8. What do you observe through the microscope? In your ScienceLog, make drawings under low and high power as you did for the onion material. Don't forget to note the magnification used. Show all detailed features as accurately as possible.

What's the Difference?

You have seen some plant cells—in the skin of an onion. You have also seen some animal cells—human cheek cells. Consult your drawings of cells, and use your memory to answer the following questions:

1. What characteristics do plant and animal cells share?

   Animal and plant cells both typically contain a cell membrane, cytoplasm, and a nucleus.

2. Would you be able to distinguish onion-skin cells from human cheek cells? If so, how would you do it? Record your thoughts here and think about this question as you carry out the next Exploration.

   Yes; the shape of an onion cell is roughly rectangular, while a cheek cell is irregularly shaped. Onion cells have a thick cell wall, while cheek cells do not.

Name _____ Date _____ Class _____

Chapter 7
Exploration Worksheet

EXPLORATION 2

What Are Plants and Animals Made Of? page 116

| Your goal | to observe, draw, and distinguish between plant and animal cells |

Safety Alert!

Iodine is an eye irritant and is somewhat corrosive to the skin. Exercise caution while using it.

You Will Need
- a microscope
- microscope slides
- coverslips
- water
- a knife
- eyedroppers
- one-quarter of an onion
- forceps
- iodine solution
- a prepared slide of human cheek cells

What to Do

1. Make a wet mount of a piece of onion skin.
   a. Remove the outside layer of one-quarter of an onion.
   b. Remove the thin skin from the inside of the layer.
   c. Cut off a small piece for your slide.
   d. Place the onion skin on a slide, add a drop of water, and place a coverslip on top.

2. Adjust the diaphragm and the light source to get the best lighting. Observe the onion skin under low power and then high power.

3. Prepare a second wet mount, but this time add a drop of iodine solution to the onion skin. How does this change what you see?
   The iodine stains certain parts of the cell, making them easier to see.

4. As you look at the onion skin, do you observe a pattern of very small parts that all look very much alike? These are the cells. In your ScienceLog, draw two or three onion-skin cells under low power and then under high power. Make note of the magnification used each time.

Illustration also on page 116 of your textbook

Exploration 3

Edible Cells, page 118

Your goal to observe, draw, and distinguish between plant and animal cells

Safety Alert!

Part 1: Plant Cells

You Will Need
- a microscope
- microscope slides
- coverslips
- a knife
- 2 straight pins
- iodine solution
- water
- eyedroppers
- pieces of plant material: potato, lettuce, green pepper, banana peel, orange peel, and carrot (optional: tea, nutmeg, pepper, mustard, coffee, and ginger)

What to Do

1. Cut or scrape off sections as thin as possible from the potato and from the banana and orange peels. Try to include a piece of skin along with scrapings of the flesh of the potato. Use the straight pins to pull the material apart. If your specimens are too thick, you may find that high power does not show details well.

2. Mount each type of plant on two slides. Use only water to mount one slide, and use a drop of iodine as stain on the other.

3. In your ScienceLog, draw the cells you observe, and label each drawing.

4. Clean up your lab area and wash your hands after you are finished.

Questions

1. Did you observe any round objects (other than air bubbles)? What do you think these might be? Of what use might they be to the plants?

   Nuclei, vacuoles, chloroplasts, and other round structures observed serve to direct the cell's activities and to store or make food.

2. When the iodine was used, did you notice a color on some slides that didn't appear on the other slides? What was the color?

   Yes; the potatoes, for instance, showed a blue to purple-black color.

3. Iodine is used as a test for the presence of starch, one of the substances in foods. Adding iodine to starch results in a blue to purple-black color. Some of the substances you examined do contain starch particles. Which plants observed gave a positive test for starch?

   The potato should test positive for starch.

4. Can you think of other sources of starch in your diet?

   Other sources of starch include pasta, rice, and oatmeal.

Exploration 3 Worksheet, continued

Part 2: Animal Cells

You Will Need
- a tiny piece of beef liver
- the same equipment used in Part 1

What to Do

1. Put a tiny piece of liver on a slide.
2. Pull apart the piece of liver using two straight pins so that the material is as thin as possible.
3. Make a wet mount. Carefully examine the sample under low and high power.
4. Add a drop of iodine to a second piece of liver as a stain, and observe this slide under high and low power as well.
5. In your ScienceLog, make two drawings of a liver cell, one with iodine and one without. Show as much detail as you can.

Questions

1. Now that you have observed cells closely, write a two-sentence answer to the question, "What are cells like?"

   Sample answer: Cells are like tiny compartments. Both plant and animal cells have nuclei and are enclosed by a cell membrane.

2. In Exploration 2, you compared plant and animal cells. Write any further observations you have that would help you distinguish between plant and animal cells.

   Plant cells have a rigid cell wall. The shape of plant cells is roughly polygonal, but animal cells have various shapes. Plant cells are often green.

Plant or Animal?

Look at the photos on page 119 of your textbook. Sort them into plant or plant-like cells and animal or animal-like cells. In your ScienceLog, state your reasons for your decisions. (Hint: Pay attention to the outer boundary of each cell. Does it appear rigid, thick, and definite in shape or delicate and flexible?)

Plant or plant-like	Animal or animal-like
a. spyrogyra, b. diatom, c. vascular tissue, d. pollen grains, i. elodea, j. root-tip cells	b. diatom, e. muscle, f. nerve cell, g. red blood cells, h. amoeba, k. paramecium

Name _____ Date _____ Class _____

Challenge Your Thinking, page 120

Chapter 7 Review Worksheet

1. A Cellular Questionnaire

Quiz yourself with the following questions:

a. How are the cells that make up living things like bricks?

Cells are like bricks in that they form a structure. Many bricks can be joined together to form a wall; likewise, many cells can be joined together to form a tissue. Cell walls of plants are rigid and give support to the organism, just as bricks give support to a building.

b. How are cells not like bricks?

Cells are unlike bricks in that each cell is made up of even smaller structures that carry out functions for the cell. These cell functions are the activities of life.

c. How do microscopes aid in the study of living things?

Microscopes are important because they allow us to examine the tiny structures that make up an organism.

d. Describe how things look different under a microscope (other than just looking larger).

The objects are also inverted.

e. Suppose that you are looking through a microscope at a tiny animal. It swims up and to the right and moves out of your field of view. Which way do you move the slide to follow it?

Down and to the left

Name _____ Date _____ Class _____

Chapter 7 Review Worksheet, continued

2. Microscope Champions

Use the table below to rate yourself as a microscope user.

Activity	Good	Fair	Could be better
a. I carry the microscope carefully to my work area.			
b. I begin observing with the lowest power objective lens.			
c. I carefully adjust the light level.			
d. I avoid touching the objective lens to the slide I am viewing.			
e. I focus the microscope by first using the coarse-adjustment knob and then using the fine-adjustment knob.			
f. I properly prepare the microscope for storage when my work is finished.			

Answers will vary. (If students rate themselves as "Fair" or "Could be better," have them write a few sentences in their ScienceLog describing how they could improve their laboratory activities.)

3. Look and See

Examine the photograph of the cell below.

Chloroplast — Nucleus — Vacuole — Cell membrane — Cell wall

Is it a plant or an animal cell? How can you tell?

The cell shown is a plant cell. The rigid, thick boundary is a cell wall, which is only found in plants.

Use the following words to label your drawing: *cell wall*, *vacuole*, *chloroplast*, and *nucleus*. Where would the cell membrane be located if it were visible? Label this also.

Name _____ Date _____ Class _____

Chapter 7
Assessment

Word Usage

1. For each group of words below, write one or more sentences. Use all of the words in the group to show how they are related.

   a. nucleus, support, cell wall, separates, activities

   Sample answer: The thick *cell wall* provides *support* and shape to the cell and *separates* individual cells. The *nucleus* controls the life *activities* of the cell.

   b. stage clips, slide, low-power lens, coarse-adjustment knob

   Sample answer: The *coarse-adjustment knob* moves the *low-power lens* down close to the stage, where a *slide* is held in place by *stage clips*.

Correction/Completion

2. The statements below are incorrect or incomplete. Your challenge is to make them correct and complete.

   a. The part of a microscope that controls the amount of light passing through an object is the eyepiece.

   The part of a microscope that controls the amount of light passing through an object is the *diaphragm*.

   b. All cells have both a cell membrane and a cell wall.

   ***Plant* cells have both a cell membrane and a cell wall, while *animal* cells have only a cell membrane.**

Numerical Problem

3. Tami is using a microscope that has the powers listed below.

   | Eyepiece = 10× | Low-power lens = 4× |
   | Medium-power lens = 10× | High-power lens = 46× |

   How many times larger will objects appear if she uses the low-power lens? the high-power lens?

   Low-power lens: 4 × 10 = 40 times larger

   High-power lens: 46 × 10 = 460 times larger

SCIENCEPLUS • LEVEL GREEN 57

Name _____ Date _____ Class _____

Chapter 7 Review Worksheet, continued

4. **Mystery Word**

   Fill in the correct "cellular" words below to find out what this unit is all about.

   C Y T O P L A S M   living matter found outside the nucleus of the cell

   V A C U O L E   space used for storage (larger and more common in plant than in animal cells)

   C H L O R O P L A S T   green, often oval, structure in plant cells; used in food production

   M I T O C H O N D R I A   oval bodies containing folded membrane; sites of cell activity where energy is released

   C E L L   W A L L   thick outer covering of the cell; provides strength and protection

   C E L L   M E M B R A N E   outer edge of an animal's cell

   N U C L E U S   dense, round or oval body; controls the life activities of the cell

   C E L L S   building blocks of all living things

56 UNIT 2 • PATTERNS OF LIVING THINGS

Name _____ Date _____ Class _____

Unit 2
Unit Activity Worksheet

A Magic Square

Complete this activity as you conclude Unit 2.

Match the items in list I with their descriptions in list II. Enter the number of the description in the box of the magic square containing the letter of the appropriate item. Add the horizontal rows, vertical rows, and diagonals to discover the MAGIC.

List I
A. snail
B. variable
C. tree rings
D. adaptation
E. positive response
F. regeneration
G. earthworm
H. butterfly
I. cancer
J. cells
K. dormancy
L. migration
M. germination
N. polar bear
O. spider
P. stimulus

List II
2. Growing roots, stems, and leaves from a seed
3. Moving to a new location in the fall or spring
4. An indication of age
5. A crab growing a new claw
6. Period when the activities of life slow down
7. A warmblooded animal
8. Motion toward light or touch
9. A change in a living thing that occurs as a result of a change in the environment
10. A condition that is changed in an experiment
11. Has segments with bristles
12. Touching the leaflets of a mimosa plant
13. Uncontrolled, harmful growth
15. An animal with six legs
16. An animal with one foot
17. Small "bricks" of living things
18. An animal with eight legs

A	B	C	D
16	10	4	9
E	F	G	H
8	5	11	15
I	J	K	L
13	17	6	3
M	N	O	P
2	7	18	12

Magic number: __39__

Name _____ Date _____ Class _____

Chapter 7 Assessment, continued

Illustration for Interpretation

CHALLENGE 1

4. Below is a good drawing of a piece of onion skin viewed under a microscope. Why is it a good drawing? List at least four reasons.

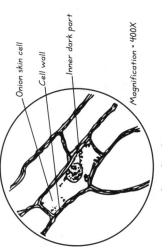

Onion skin cell
Cell wall
Inner dark part
Magnification = 400X
Onion Skin Cell

Answers could include the following: A sharp, hard pencil was used on plain paper; details are shown; only one cell is drawn, with all of the observed details; labels are outside the drawing and printed neatly; label lines end at the feature being identified; there is no shading, blurring, or added color; the magnification is included; the drawing was done neatly and carefully.

Short Response

CHALLENGE 2

5. Why don't animal cells have cell walls?

Sample answer: Cell walls are the structures that make plants rigid. Animals are more mobile than plants and must, therefore, have more flexible parts. The flexibility of the cell membrane allows for flexibility in larger structures.

Unit 2
Unit Review Worksheet

Making Connections, page 122

The Big Ideas

In your ScienceLog, write a summary of this unit, using the following questions as a guide:

1. What are some signs of life? (Ch. 4)
2. How do plant and animal movement differ? (Ch. 4)
3. In what ways are the growth patterns of plants and people similar? (Ch. 5)
4. In what ways are they different? (Ch. 5)
5. What are some of the different forms of growth? (Ch. 5)
6. What is the importance of each? (Ch. 5)
7. What does "response to stimuli" mean? (Ch. 6)
8. Why is this response valuable? (Ch. 6)
9. What evidence is there that animals have a sense of time? (Ch. 6)
10. What are some adaptations of plants and animals over long periods of time? (Ch. 6)
11. How do plant and animal species adapt over long periods of time? (Ch. 6)
12. How are plant and animal cells similar? How do they differ? (Ch. 7)
13. Describe the procedure for making a wet mount. (Ch. 7)

A sample unit summary is provided on page 122 of the Annotated Teacher's Edition.

Checking Your Understanding

1. Would a seed be an example of a living thing? Why or why not?

   A seed is a living plant embryo that is waiting for the right conditions to begin development.

2. What are some of the reasons that animals migrate? Name at least three animals that migrate.

   Reasons for migrating include avoiding unfavorable weather and the need to find food or resources. Three types of animals that migrate are whales, bats, and birds.

Unit 2 Review Worksheet, continued

3. Look at the illustration below showing the life cycle of a butterfly. How does the butterfly's growth pattern differ from that of a human?

   The life cycle of a butterfly consists of four distinct stages—egg, larva, pupa, and adult. These stages differ in appearance, movement, and feeding habits. The growth pattern of a human, from egg to adult, is a more gradual process and does not include distinct stages like those of the butterfly.

4. **Construct a concept map** using the following terms: *ectothermic, bear, endothermic, animals,* and *frog.*

Sample concept map:

Animals can be endothermic such as a bear

Animals can be ectothermic such as a frog

Unit 2 Review Worksheet, continued

5. What are some of the different ways in which animals move?
   Crawling, walking, swimming, and flying

6. Are all forms of growth useful? Explain.
   Uncontrolled growth is not useful. Cancerous growth, for instance, can be harmful or even fatal.

7. Is bigger always better? Explain.
   Large size has its disadvantages. Larger organisms, such as the African elephant, must spend more time finding and eating food than do their smaller counterparts, for instance.

8. How do snakes demonstrate adaptation to their environment?
   The scales of a snake help the animal grip the ground so that it can push itself forward.

9. In your own words, compare an earthworm's locomotion to that of a snake.
   A snake moves either by flexing its body in an S-shaped curve or by using its scales to move forward in a straight line. An earthworm propels itself forward by alternately contracting and extending its muscles. The muscles move the earthworm's bristles, which push the earthworm along.

10. How might a sense of time be helpful to animals? to plants?
    A sense of time in animals and plants helps them survive the difficulties of their environments. For example, animals might sleep during the day to avoid predators or migrate in the winter to find food. Some plants close up their leaves at night to conserve heat and water or drop their leaves in the fall to reduce the chance of freezing in the winter.

Unit 2 Review Worksheet, continued

11. Add the following to your list of "Curious Questions." Suggest answers to each.

    a. Where do snakes, frogs, and salamanders go in the winter?
       Some snakes, frogs, and salamanders become dormant in the winter. They may spend the winter underground.

    b. How do endothermic animals keep their body temperature constant?
       Endothermic animals use the energy in food to regulate temperature. Mammals have a layer of fat under their skin that serves as insulation. Animals with hair or feathers can fluff up this outer covering to trap air and insulate themselves.

    c. What advantages do endothermic animals have over ectothermic animals?
       Endothermic animals can typically function normally even when their environment is cold. They can also live in areas that experience extreme temperatures, whereas ectothermic animals may not be able to live there.

    d. What causes animals to start migrating? to stop migrating?
       Environmental cues such as day length may cause animals to start migrating. A combination of learned and innate behavior may signal the end of migration.

    e. What cues do animals use to guide them as they migrate?
       Cues from the environment include landmarks, wind direction, and the position of the sun.

Name _____ Date _____ Class _____

Unit 2
End-of-Unit Assessment

Word Usage

1. For each group of words below, write one or two sentences. Use all of the words in the group to show how they are related.

   a. stimulus, respond, worm, plant, light

   Sample answer: A *worm* and a *plant respond* differently to the same *light stimulus*. The plant moves toward the light, and the worm moves away from it.

   b. cooling system, sweating, panting, dog

   Sample answer: The *cooling system* for a human is *sweating*. The *cooling system* for a *dog* is *panting*.

Correction/ Completion

2. The statements below are either incorrect or incomplete. Make the statements correct and complete.

   a. When a tree is cut, growth rings may be visible. The number of rings tells the tree's __**age**__, and the width of a ring _____ that year.

   b. After a lobster loses a claw in a fight, the claw may __**grow back**__. This is an example of regeneration.

   c. Humpback whales are dormant in winter; they travel south to warmer waters.

   Humpback whales *migrate* in winter; they travel south to warmer waters.

Short Response

3. Tell how a cactus plant is adapted to its environment.

   Sample answer: A cactus plant is adapted to its environment because it has an extensive root system to collect water, a waxy trunk to store water, and spiny leaves to prevent water loss and to protect the plant.

Name _____ Date _____ Class _____

Unit 2 Assessment, continued

Graph for Correction or Completion

Sample completed graph:

4. A bean plant is watered regularly for 4 weeks. After that, it is no longer watered. Show what the graph might look like after the fourth week.

Illustrations for Correction or Completion

5. Complete the drawing by showing how the plant in the illustration would look 3 weeks after being moved away from the light.

Before / After

The drawing should show the plant leaning toward the light.

Name _____ Date _____ Class _____

Unit 2 Assessment, continued

6. Complete the drawing by showing the location of the worm in the box after more light has been allowed into the box.

Before / After

The drawing should show that the worm has moved farther toward the right corner.

Numerical Problem

7. **The Great Worm Race Between Alpha, Beta, and Gamma**
The three worms started from the center of the ring. Alpha sped outward at 50 cm/h for the first half-hour of the one-hour race. During the last half-hour, his speed was only 30 cm/h. Beta, on the other hand, started out slowly; during the first half-hour her speed was only 20 cm/h. However, in the final half-hour, she sped up to 70 cm/h. Gamma had a steady pace for the full hour, averaging 40 cm/h. Who went the farthest in 1 hour and won the race? How far did the winner go? How far did the losers go? Show your work.

Beta won the race by going 45 cm in 1 hour. The losers each went

40 cm.

Short Essay

8. Describe an imaginary, migratory animal. Identify the path of its migration, when it migrates, and what factors cause it to migrate.

Sample answer: The needle-nosed warbler is a small, migratory bird that spends summers along the shores of the Great Lakes and winters in southern parts of Texas and Louisiana. In the spring, as the days grow longer and the temperature rises, the birds' internal clocks alert them that it is time to migrate northward. As the temperature cools in the northern United States, the birds respond by flying south. By migrating, they are able to find food all year long, and they avoid weather that is too cold or too warm for them to survive.

Name _____ Date _____ Class _____

Unit 2 Assessment, continued

Illustration for Interpretation

9. Study the illustration below showing the life cycle of a frog.

a. Which features of the tadpole are adaptations to life in the water?
   Suction disk, gills, tail

b. How does the tadpole move?
   By moving its tail

c. Which features of the frog are adaptations to life on land?
   Front and back legs, lungs

d. How does the frog move on land?
   By hopping on its back two legs

e. Why is water important to the frog's reproduction?
   The eggs are laid and fertilized in the water, and the offspring develop in the water. Without water, frogs could not successfully reproduce.

Name _____ Date _____ Class _____

Unit 2 Assessment, continued

Data for Interpretation

10. Consider this table of students' heights.

Average Height (cm)

Grade	Boys	Girls	Raj	Marilyn
6	125	125	115	138
7	128	130	118	153
8	132	136	122	166
9	138	143	126	170
10	148	149	135	171
11	161	154	155	172
12	172	158	173	172

a. During which grade did the average height for girls increase the most? __Grade 9__

b. During which grade did the average height for boys increase the most? __Grade 11__

c. During which grade did Raj grow the most? __Grade 11__

d. During which grade did Marilyn grow the most? __Grade 7__

e. Using the data above for boys and girls, complete the bar graph below.

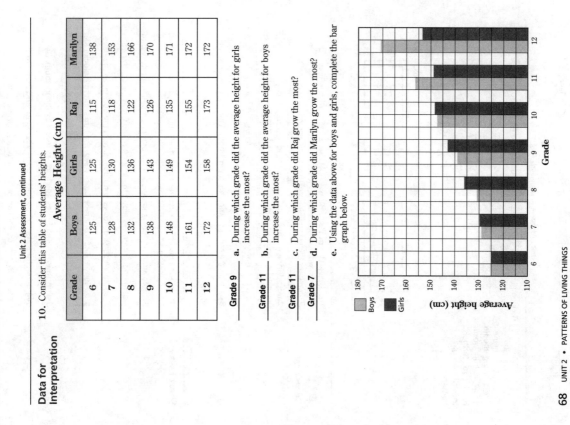

Name _____ Date _____ Class _____

Activity Assessment, continued

Data Chart

Material, substance, or organism	Observations
Answers will vary depending on samples but should be clear and complete.	

On a separate piece of paper, create a classification system summarizing the information in this chart.

Name _____ Date _____ Class _____

Unit 2 SourceBook Review Worksheet

Unit CheckUp, page S63

Concept Mapping

The concept map below illustrates major ideas in this unit. Complete the map by supplying the missing terms. Then extend your map by answering the additional question beneath it.

Sample concept map:

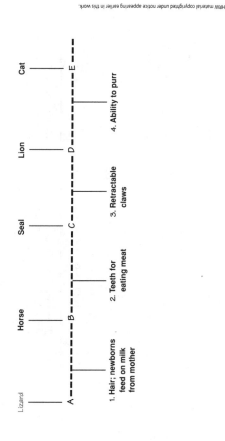

Checking Your Understanding

Where and how would you connect the terms *DNA* and *organelles*?

Select the choice that most completely and correctly answers the following questions.

1. Starfish and humans are members of the same
   a. (kingdom). b. family. c. genus. d. species.
2. The material that transmits genetic instructions between generations is
   a. mitochondria. b. cytoplasm. c. (DNA). d. testosterone.

Name _____ Date _____ Class _____

SourceBook Activity Worksheet, continued

1. What characteristics do *all* of the animals except the lizard share? Write these characteristics in the space labeled *1* on the cladogram. This point marks the point where all of the other animals split from their common ancestor, the lizard. The other animals occupy a different branch on the cladogram because they have adapted to specific environmental conditions.

2. What characteristic is shared by only *three* of the animals? Write this characteristic in the space labeled *2* on the cladogram.

3. What characteristic is shared by only *two* of the animals? Write this characteristic in the space labeled *3* on the cladogram.

4. What characteristic is exhibited by just *one* of the animals? Write this characteristic in the space labeled *4* on the cladogram.

5. Now classify the animals. To do this, write the name of each animal on the correct branch of the cladogram. (An animal must exhibit all of the characteristics to the left of its branch.) The lizard is already shown on branch *A*. Remember that at least one unique characteristic separates the animals on each branch of the cladogram.

Lizard   Horse   Seal   Lion   Cat

A — B — — — C — — — D — — — E

1. Hair; newborns feed on milk from mother
2. Teeth for eating meat
3. Retractable claws
4. Ability to purr

SourceBook Review Worksheet, continued

Critical Thinking

Carefully consider the following questions, and write a response that indicates your understanding of science.

1. On a late-night television show, a self-proclaimed psychic predicts that four entirely new biological kingdoms will be discovered in the coming year. Is this a realistic prediction? Explain.

   No; it is not realistic. Kingdom is such a basic biological category that it is unlikely that organisms so fundamentally different from those already known could have been overlooked.

2. In a certain population of foxes, 95 percent are reddish in color, and the rest are solid white. Foxes feed by hunting small animals. Suppose that the climate (which is now fairly moderate with little winter snow) changes and becomes cold and snowy for much of the year. How might this affect the ratio mentioned in the first sentence of this question? Explain.

   The ratio would probably change. As predators, red foxes would be at a severe disadvantage for much of the year because of their coloration, which would stand out against the white snow, thus making it difficult to sneak up on prey. White foxes would not face this problem.

3. Certain microorganisms carry out photosynthesis but are also able to take in food from their surroundings and to move about on their own. Speculate how the discovery of organisms such as these helped to do away with the two-kingdom system of classification.

   Such organisms have traits normally found either in plants or in animals, but not both, thus pointing out the need for additional classification categories.

SourceBook Review Worksheet, continued

3. Which of the following is NOT a function of the human skin?
   a. waste removal
   b. temperature regulation
   c. (gas exchange)
   d. protection

4. The body system enabling you to read and understand this question is the
   a. optic system.
   b. (nervous system.)
   c. immune system.
   d. educational system.

5. Chromosomes can be compared with
   a. ladders.
   b. security fences.
   c. a shop floor.
   d. (blueprints.)

6. Which choice correctly lists the levels of organization from simplest to most complex?
   a. organism, organ system, organ, tissue, cell
   b. (cell, tissue, organ, organ system, organism)
   c. tissue, organ system, organism, cell, organ
   d. organ, organ system, organism, tissue, cell

Interpreting Illustrations

The accompanying illustration compares the forelimbs of a whale and a human. Identify the corresponding parts, and describe below how these similarities point to an evolutionary relationship between whales and humans.

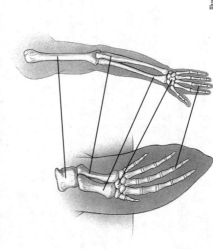

Illustration also on page S64 of your textbook

Because there are so many corresponding parts, it is unlikely that the forelimbs of these two organisms could have evolved independently.

Name _____ Date _____ Class _____

Unit 2 SourceBook Assessment

1. The jellylike fluid found within a living cell is composed mostly of
   a. carbohydrates.   b. proteins.   c. fats.   d. (water.)   e. oxygen.
2. The members of a _____ have the most characteristics in common.
   a. phylum   b. kingdom   c. (species)   d. genus
3. According to Darwin's theory of evolution, an archer fish got its ability to squirt water because it
   a. practiced it.   b. acquired it.   c. (inherited it.)
4. Which of the following is NOT one of the three parts of the cell theory?
   a. Only living cells can produce new living cells.
   b. The cell is the basic unit of all living things.
   c. (Living cells come from nonliving substances.)
5. What level of organization is represented by each of the following?

   a. tree   **Organism**
   b. heart   **Organ**
   c. blood   **Tissue**
   d. skeleton **System**

6. Identify the cell structure whose function would be most like that of the following parts of a school:
   a. the office   **Nucleus**
   b. the outer walls **Cell wall**
7. An animal that acquires a specific trait during its lifetime may pass that trait on to its offspring.
   a. true   b. (false)
8. Which two organisms would be more closely related: *Ursus arctos* and *Ursus maritimus*, or *Coriandrum sativum* and *Lepidium sativum*?
   Ursus arctos* and *Ursus maritimus
9. Discuss the main advantage that specialized cells have over unspecialized cells.
   Answers will vary but should address the fact that specialized cells work more efficiently than unspecialized cells.

SCIENCEPLUS • LEVEL GREEN   79

Name _____ Date _____ Class _____

SourceBook Review Worksheet, continued

4. How does the cell wall determine some of the major defining characteristics of plants? Why would a cell wall be a disadvantage in animal cells?
   The rigid cell wall provides the plant with structural support. This rigidity would be a disadvantage for animal tissues such as muscles or nerves, which must be very supple.

Portfolio Idea In your ScienceLog, write a story in which you are an organ system, and describe a typical day in your life. Describe what you do, why you do it, how you work with other organ systems, typical problems that might affect you, and so on. Be creative but factually accurate.
   Answers will vary, but should reflect an understanding of the form and function of the organ system that each student chooses. Accept all reasonable responses.

78 UNIT 2 • PATTERNS OF LIVING THINGS

SourceBook Assessment, continued

10. Can some organs perform more than one function? If so, give an example of an organ with more than one function, and list its functions.
    Yes. The testes and the ovaries, which produce both reproductive cells and hormones, are two possible answers.

11. Correct this statement: The proportion of genes you receive from each of your parents depends on which parent's genes were dominant.
    The proportion of genes you receive from each of your parents *is the same; half comes from one parent and half from the other.*

12. What is the function of chlorophyll in plant cells?
    Chlorophyll allows plant cells to make their own food through the process of photosynthesis.

13. List three things that scientists might study to classify a newly discovered life-form.
    Answers will vary but may include appearance, structure, fossils, biochemical makeup, reproductive habits, and DNA.

14. Classify the following items into two groups: paper clip, apple, pencil, paper, staples, and granola bar. Then explain the characteristics you used to classify the items.
    Answers will vary. Sample answer: group 1—paper clip, pencil, paper, staples; group 2—apple, granola bar. Group 1 consists of objects found in a school or office, and group 2 consists of types of food.

SourceBook Assessment, continued

15. Why is the scientific system of naming living things necessary? If so, give an example of an organ with more than one function, and list its functions. What would biology be like without it?
    Scientific names are important because they are universally established and mean the same thing to everyone, regardless of native language or dialect. This naming system allows scientists to discuss with clarity how animals are related. Without this system, biology would probably be confusing and extremely complicated.

16. Would it be possible for someone to believe in both the theory of spontaneous generation and the cell theory? Explain your answer.
    No. Spontaneous generation is the theory that life arises from non-living materials, and the third step of the cell theory states that "only living cells can produce new living cells."

17. Why would you expect to find a lot of mitochondria in a muscle cell?
    Muscle cells should have a lot of mitochondria because these are the organelles responsible for cellular respiration, which occurs more often in active cells, such as those in muscles.

18. Match the kingdom on the left with the correct characteristics on the right.

    a. Animalia — **d** — All make food and have cell walls.
    b. Fungi — **a** — All are multicellular and ingest food.
    c. Monera — **c** — Cells do not have nuclei.
    d. Plantae — **b** — All have nuclei and must absorb food from surroundings.
    e. Protista — **e** — None have organization above tissue level.

SourceBook Assessment, continued

19. Match the cell structures on the left with the correct function on the right.

- **a.** chromosome — _a_ contains instructions for cell processes
- **b.** cytoplasm — _e_ contains DNA and RNA and controls the cell
- **c.** membrane — _c_ limits access to cell and aids in gathering food
- **d.** mitochondria — _b_ place where instructions are carried out
- **e.** nucleus — _f_ protein-manufacturing site
- **f.** ribosome — _g_ stores food and waste products
- **g.** vacuole — _d_ turns sugar into energy by cellular respiration

20. Match each element with its corresponding organ system.

- **a.** sensory neurons — _c_ circulatory system
- **b.** bicep — _d_ skeletal system
- **c.** aorta — _f_ integumentary system
- **d.** marrow — _b_ muscular system
- **e.** kidneys — _a_ nervous system
- **f.** epidermis — _h_ digestive system
- **g.** alveoli — _e_ excretory system
- **h.** gallbladder — _g_ respiratory system

SourceBook Assessment, continued

21. Complete the following chart to show the similarities and differences between the cellular structure of plants and animals.

Sample chart:

Plant cell	Animal cell
cell membrane	cell membrane
cell wall	
nucleus	nucleus
chromosomes	chromosomes
chloroplasts	
large vacuoles	vacuoles
ribosomes	ribosomes
DNA	DNA
mitochondria	mitochondria

22. Describe what would happen if an earthquake suddenly created a giant gorge, dividing a population of wolves.

Sample answer: Because there would be two new populations of wolves that no longer interbred or had any contact with each other, they would begin to develop differently. Variations among the two populations would arise independently and begin to alter their genetic make-up. Eventually, over a long period of time, the two populations would evolve into two different species.

23. Give two examples, one from a plant and one from an animal, of a tissue, an organ, and a system.

Answers will vary. Sample answer: An example of a plant tissue is the epidermal tissue. The transport system consists of three organs: the roots, the stems, and the leaves. An example of an animal tissue is epithelial tissue. The integumentary system consists of the skin, the largest human organ.

Name _____ Date _____ Class _____

SourceBook Assessment, continued

24. Describe the different types of muscles that make up the muscular system.

Sample answer: Smooth muscles, which are found in places like the digestive tract, consist of sheets of tissue. They are responsible for involuntary movement. Skeletal muscles are voluntarily controlled and are found in places like the arms and legs and are generally attached to bones. Cardiac muscles are the most active muscles, and they are found only in the heart. They contain a lot of mitochondria and move involuntarily.

25. Explain how human bodies fight diseases internally, and explain why people usually suffer from chickenpox only once.

Sample answer: When a foreign substance, such as a virus or microorganism, enters the circulatory system, white blood cells can create an antibody to destroy it. The white blood cells then store the information about how to make the antibody in case that particular substance invades the body again. People usually suffer from chickenpox only once because their white blood cells "remember" how to make the antibody to the foreign substance that causes the illness. This resistance to disease is known as acquired immunity.